STORM WATER DISCHARGES

Regulatory Compliance and Best Management Practices

Mark S. Dennison

CRC

LEWIS PUBLISHERS

Boca Raton New York London Tokyo

Library of Congress Cataloging-in-Publication Data

Dennison, Mark S.
Storm water discharges : regulatory compliance and best management practices / by Mark Dennison
p. cm.
Includes bibliographical references and index.
ISBN 1-56670-198-8 (alk. paper)
1. Runoff—Management. 2. Storm sewers—Law and legislation—United States. 3. Environmental permits—United States.
I. Title.
TD675.D45 1996
363.73'94—dc20 95-40473
CIP

Direct all inquiries to CRC Press, Inc., 2000 Corporate Blvd., N.W., Boca Raton, Florida 33431.

Lewis Publishers is an imprint of CRC Press

International Standard Book Number 1-56670-198-8
Library of Congress Card Number 95-40473
Printed in the United States of America 1 2 3 4 5 6 7 8 9 0
Printed on acid-free paper

For Tracey, the love of my lifetime

About the Author

Mark Dennison is an attorney and author or co-author of numerous books and articles dealing with environmental law and regulatory compliance issues. His books include *Environmental Reporting, Recordkeeping, and Inspections: Compliance Guide for Business and Industry* (1995); *Environmental Due Diligence for Lenders: Practical Guide to Implementing an Environmental Risk Program*, with R. Kenneth Keim and Philip Lee (1995); *Pollution Prevention Strategies and Technologies* (1995); *OSHA and EPA Process Safety Management Requirements: A Practical Guide for Compliance* (1994); *Hazardous Waste Regulation Handbook: A Practical Guide to RCRA and Superfund* (1994); *Understanding Solid and Hazardous Waste Identification and Classification* (1993); *Wetlands and Coastal Zone Regulation and Compliance,* with Steven Silverberg (1993); *Wetlands: Guide to Science, Law, and Technology*, with James Berry (1993).

Mr. Dennison is also the editor-in-chief of the monthly newsletter, *Environmental Strategies for Real Estate*. He is in private practice in Ridgewood, New Jersey, where he specializes in environmental, land use, real estate, and zoning law. He is admitted to practice in New Jersey and New York. Mr. Dennison holds a B.A., *magna cum laude*, from the State University of New York (Oswego), an M.A. from Syracuse University, and a J.D. from New York Law School.

Acknowledgments

I wish to express my appreciation to everyone who played a role in the publication of this book. I thank the editorial and production staff at CRC Press/Lewis Publishers, especially Bob Esposito, Publisher - Environmental Sciences, for inviting me to write the book and for providing his editorial leadership throughout the publishing process.

I also wish to thank the many helpful individuals working at different federal and state regulatory offices who answered my questions and provided copies of various forms and government documents.

Additional thanks go to family and friends for providing moral support and inspiration, especially Erin Carrather, Jonathan Huff, Jessie Ann Huff, Bobbie Waits, Harry Huff, Jr., Mrs. J. Howard Gould, and Keith Dennison.

But, as always, my greatest thanks must go to my dear Tracey, whose constant love and support provided me with the inspiration necessary to complete this book.

M.S.D.

Preface

Storm Water Discharges: Regulatory Compliance and Best Management Practices offers a practical, nuts-and-bolts approach to compliance with EPA's complex Storm water discharge requirements. This useful guide has been written in non-lawyer language, explaining the essential regulatory issues in a way that can be directly applied to a company's day-to-day operations. The book contains useful tables summarizing key points, helpful checklists, and valuable appendices, including a sample Storm Water Pollution Prevention Plan.

In addition to a step-by-step discussion of the regulatory requirements, the book also focuses on the technological aspects of complying with the storm water discharge regulations. The book provides in-depth coverage of the development of storm water pollution prevention plans along with different types of site-specific practices needed to implement the plans. The book offers comprehensive treatment of Best Management Practices for different types of facilities and activities to aid in eliminating or reducing potential storm water contamination.

Mark S. Dennison
Ridgewood, New Jersey

Table of Contents

Chapter 6
INDUSTRY-SPECIFIC BMPs 217

Chapter 7
ACTIVITY-SPECIFIC BMPs 247

Chapter 1

NATURE OF STORM WATER POLLUTION

1.1 SCOPE OF THE STORM WATER PROBLEM

The discharge of storm water into navigable waterways is recognized as a widespread problem responsible for much of the degradation of U.S. waterways. According to the U.S. Environmental Protection Agency (EPA), "pollution from diffuse sources, such as runoff from agricultural, urban areas, construction sites, land disposal and resource extraction, is cited by the States as the leading cause of water quality impairment."[1] The problem is difficult to solve primarily because storm water pollution is easily created: a discharge occurs whenever rainwater falls on contaminated soils or piles of materials containing pollutants; this storm water then carries the contaminants into waterways.

1.1.1 National Assessments of Storm Water

Several national assessments have been conducted to evaluate the impacts of diffuse sources of storm water runoff on receiving water quality. The *National Water Quality Inventory, 1990 Report to Congress* provides a general assessment of water quality based on biennial reports submitted by the States under Section 305(b) of the Clean Water Act (CWA). In the Section 305(b) Reports, States indicate the fraction of the States' waters that have been assessed, the fraction of those assessed waters that are not supporting designated uses, and the sources of use impairment for those waters (e.g., diffuse sources, point sources, and natural sources). The 1990 Report to Congress indicates that roughly 30 to 40% of assessed rivers, lakes and estuaries are not supporting the uses for which they are designated. Based on information from the 51 States and Territories that reported on sources of pollution, the 1990 Report indicates that storm water runoff from a number of diffuse sources, including agricultural areas, urban areas, construction sites, land disposal activities, and resource extraction activities, is the leading cause of water quality impairment cited by States. For those States reporting in each category, diffuse sources were cited as causing use impairments in the following magnitudes: For rivers and streams, 11% of impaired river miles are caused by separate storm sewers, 6% are caused by construction activities, and 14% are caused by resource extraction. For lakes, 28% of impaired lake acres are caused by separate storm sewers and 25% are caused by land disposal. For the Great Lakes' shoreline, 6% of impaired shoreline miles are caused by separate storm sewers, and 41% are caused by land disposal. For estuaries, 30% of

[1] National Pollution Discharge Elimination System Permit Application Regulations for Storm Water Discharges, 55 Fed. Reg. 47,990, 47,991 (Nov. 16, 1990).

impaired acres are caused by separate storm sewers. For coastal areas, 36% of impairments are caused by separate storm sewers, and 37% are caused by land disposal.

1.1.2 Storm Water Discharge Pollutants Associated with Industrial Activity

The volume and quality of storm water discharges associated with industrial activity depend on a number of factors, including the industrial activities occurring at the facility, the nature of precipitation, and the degree of surface imperviousness. Rainwater may pick up pollutants from structures and other surfaces as it drains from the land. In addition, sources of pollutants other than storm water, such as illicit connections,[2] spills, and other improperly dumped materials, may increase the pollutant loads discharged from separate storm sewers. The sources of pollutants in storm water discharges differ with the type of industry operation and specific facility features. For example, air emissions may be a significant source of pollutants at some facilities, material storage operations at others, and still other facilities may discharge storm water associated with industrial activity with relatively low levels of pollutants.

Studies have shown that many storm sewers contain illicit nonstorm water discharges and that large amounts of wastes are disposed of improperly in storm sewers.[3] Removal of these discharges presents opportunities for dramatic improvements in the quality of storm water discharges. Storm water discharges from industrial facilities may contain, in addition to illicit connections and improperly disposed wastes, toxics and conventional pollutants when material management practices allow exposure to storm water.

In some municipalities, illicit connections of sanitary, commercial, and industrial discharges to storm sewer systems

---

[2] Illicit connections are contributions of unpermitted nonstorm water discharges to storm sewers from any of a number of sources including sanitary sewers, industrial facilities, commercial establishments, or residential dwellings.

[3] See, e.g., Final Report, Nationwide Urban Runoff Program, Vol. 1 (EPA, 1983).

have had a significant impact on the water quality of receiving waters. Studies have shown that illicit connections to storm sewers can create severe, widespread contamination problems.[4]

Six activities can be identified as major potential sources of pollutants in storm water discharges associated with industrial activity:

1. Loading or unloading of dry bulk materials or liquids;
2. Outdoor storage of raw materials or products;
3. Outdoor process activities;
4. Dust or particulate generating processes;
5. Illicit connections or inappropriate management practices; and
6. Waste disposal practices.[5]

The potential for pollution from many of these activities may be influenced by the presence and use of toxic chemicals.

Loading or unloading of dry bulk materials or liquids. Loading and unloading operations typically are performed along facility access roads and railways and at loading/unloading docks and terminals. These operations include pumping of liquids or gases from truck or rail car to a storage facility or vice versa,

---

[4] For example, the Huron River Pollution Abatement Program inspected 660 businesses, homes, and other buildings located in Washtenaw County, Michigan, and found that 14% of the buildings had improper storm drain connections. Illicit discharges were detected at a higher rate of 60% for automobile-related businesses, including service stations, automobile dealerships, car washes, body shops, and light industrial facilities. While some of the problems discovered in this study were due to improper plumbing or illegal connections, a majority were approved connections at the time they were built, but have since become unlawful discharges. 57 Fed. Reg. 41236, 41238 (Sept. 9, 1992).

[5] See, e.g., The Cadmus Group, Manual of Practice: On Site Identification of Illicit Connections (1990); Design of Urban Runoff Quality Controls, American Society of Civil Engineers (1988); Torno, Urban Storm water Quality Enhancement - Source Control, Retrofitting, and Combined Sewer Technology, American Society of Civil Engineers (1989); Analysis of Implementing Permitting Activities for Storm Water Discharges Associated with Industrial Activity, U.S. EPA (1991).

pneumatic transfer of dry chemicals to or from the loading or unloading vehicle, transfer by mechanical conveyor systems, and transfer of bags, boxes, drums, or other containers from vehicle by forklift trucks or other materials handling equipment. Material spills or losses may discharge directly to the storm drainage systems or may accumulate in soils or on surfaces and be washed away during a storm or facility washdown.

Outdoor storage of raw materials or products. Outdoor storage activities include the storage of fuels, raw materials, byproducts, intermediates, final products, and process residuals. Methods of material storage include using storage containers (e.g., drums or tanks), platforms or pads, bins, silos, boxes, or piles. Materials, containers, and material storage areas that are exposed to rainfall and/or runoff may contribute pollutants to storm water when solid materials wash off or materials dissolve into solution.

Outdoor process activities. Other outdoor activities include certain types of manufacturing and commercial operations and land-disturbing operations. Although many manufacturing activities are performed indoors, some activities, such as equipment maintenance and/or cleaning, timber processing, rock crushing, vehicle maintenance and/or cleaning, and concrete mixing, typically occur outdoors. Processing operations may result in liquid spillage and losses of material solids to the drainage system or surrounding surfaces, or creation of dusts or mists, which can be deposited locally. Some outdoor industrial activities cause substantial physical disturbance of land surfaces that result in soil erosion by storm water. For example, disturbed land occurs in construction and mining. Disturbed land may result in soil losses and other pollutant loadings associated with increased runoff rates. Facilities whose major process activities are conducted indoors may still apply chemicals such as herbicides, pesticides, and fertilizer outdoors for a variety of purposes.

Dust or particulate generating processes. Dust or particulate generating processes include industrial activities with stack emissions or process dusts that settle on plant surfaces. Localized atmospheric deposition is a particular concern with heavy manufacturing industries. For example, monitoring of areas surrounding smelting industries has shown much higher levels of metals at sites nearest the smelter. Other industrial sites, such as mines, cement manufacturing, and refractories, generate significant levels of dusts.

Illicit connections or inappropriate management practices. Illicit connections or inappropriate management practices result in improper nonstorm water discharges to storm sewer systems. The likelihood of illicit discharges to storm water collection systems is expected to be higher at older facilities, due to past practices, as well as for facilities that use high volumes of process water or dispose of significant amounts of liquid wastes, including process wastewaters, cooling waters, and rinse waters. Pollutants from nonstorm water discharges to the storm sewer system of individual facilities are caused typically by a combination of improper connections, spills, improper dumping, and the belief that the absence of visible solids in a discharge is equivalent to the absence of pollution. Illicit connections are often associated with floor drains that are connected to separate storm sewers. Rinse waters used to clean or cool objects discharge to floor drains connected to separate storm sewers. Large amounts of rinse waters may originate from industries that use regular washdown procedures; for example, bottling plants use rinse waters for removing waste products, debris, and labels. Rinse waters can be used to cool materials by dipping, washing, or spraying objects with cool water; for example, rinse water is sometimes sprayed over the final products of a metalplating facility for cooling purposes. Condensate return lines of heat exchangers often discharge to floor drains. Heat exchangers, particularly those used under stressed conditions (such as exposure to corrosive fluids) such as in the metal finishing and electroplating industry, may develop pinhole leaks that result in contamination of condensate by process wastes. These and other nonstorm water discharges to a storm sewer may be intentional, based on the belief that the discharge (condensate in the example previously discussed) does not contain pollutants, or they may be inadvertent, if the operator is unaware that a floor drain is connected to the storm sewer.

Waste disposal practices. Waste management practices include temporary storage of waste materials, operating landfills, waste piles, and land application sites that involve land disposal. Outdoor waste treatment operations also include waste water and solid waste treatment and disposal processes, such as waste pumping, additions of treatment chemicals, mixing, aeration, clarification, and solids dewatering. Facilities often conduct some waste management on site.

1.2 POLLUTANTS IN STORM WATER AND ASSOCIATED IMPACTS

The nature of the impacts associated with specific urban storm water pollutants are reviewed in this section. The following types of pollutants found in urban runoff are addressed:

* *Sediment / habitat alteration;*
* *Oxygen-demanding substances* (organic matter);
* *Nutrients* (phosphorus, nitrogen);
* *Toxic substances* (heavy metals, oil and grease);
* *Bacteria;*
* *Floatables;* and
* *Multiple impacts* of several of these pollutants acting in concert.

1.2.1 Sediment/Habitat Alteration

High concentrations of suspended sediment in streams can cause multiple impacts including increased turbidity, reduced light penetration, reduced prey capture for sight-feeding predators, clogging of gills/filters of fish and aquatic invertebrates, reduced spawning and juvenile fish survival, and reduced angling success. Additional impacts result after sediment is deposited in slower moving receiving waters, such as smothering of the benthic community, changes in the composition of the bottom substrate, more rapid filling of small impoundments which create the need for costly dredging, and reduction in aesthetic values. Sediment having a high organic or clay content is also an efficient carrier of toxicants and trace metals. Once deposited, pollutants in these enriched sediments can be remobilized under suitable environmental conditions to pose an additional risk to benthic and other aquatic life.

The greatest sediment loads are exported during the construction phase of any development activity. Furthermore, in intensively developed watersheds, increased stream flow can result in channel degradation, requiring stream bank erosion control.

1.2.2 Oxygen-Demanding Substances

Decomposition of organic matter by microorganisms depletes dissolved oxygen (DO) levels in receiving waters, especially slower moving streams and lakes and estuaries. There are several measures of the degree of potential DO depletion, the most common of which are the Biochemical Oxygen Demand (BOD_5) test and the Chemical Oxygen Demand (COD) test. Both of these tests have problems associated with their use in urban runoff, but it has been demonstrated (e.g., Rouge River, Western Long Island Sound) that urban runoff can severely depress DO levels after large storms, and that BOD_5 solids can accumulate in bottom sediment causing impacts during periods of dry weather. BOD_5 levels can exceed 10 to 20 mg/l during storm events which can lead to anoxic conditions (zero oxygen) in shallow, slow moving, or poorly flushed receiving waters. The problem is particularly acute in some older urban areas, where storm runoff BOD_5 mixes with overflows from combined or sanitary sewers.

The greatest export of BOD_5 typically occurs from older, highly impervious, highly populated urban areas with outdated combined storm sewers. In contrast, only moderate BOD_5 export has been reported from newer, low-density suburban residential development.

1.2.3 Nutrients

The levels of phosphorus and nitrogen in urban runoff can lead to accelerated eutrophication in downstream receiving waters. Generally, phosphorus is the controlling nutrient in freshwater systems. The greatest risk of eutrophication is in urban lakes and impoundments with long detention times (two weeks or greater). Surface algal scums, water discoloration, strong odors, depressed oxygen levels (as the bloom decomposes), release of toxins, and reduced palatability to aquatic consumers are among the problems encountered. High nutrient levels can also promote the growth of dense mats of green algae that attach to rocks and cobbles in shallow, unshaded headwater streams. High nutrient loads from urban runoff, in combination with other sources, can contribute to eutrophication in both fresh and tidal waters.

As a general rule of thumb, as impervious area increases, nutrients build up on surfaces and runoff transport capacities rise as well, leading to high pollution loads. Exceptions include land under development and land activities that receive unusually high fertilizer inputs, such as golf courses, cemeteries, and other intensively landscaped areas. Examples of eutrophication caused by nonpoint sources of nutrients include the Dillon Reservoir in Colorado and the Occoquan Reservoir in Virginia.

1.2.4 Toxic Substances

Toxic substances are all defined as materials capable of producing an adverse response or effect in a biological system. A large number of potentially toxic compounds are routinely detected in urban storm water. These include trace metals (lead, zinc, copper, and cadmium), pesticides and herbicides, and hydrocarbons (derived from oil and grease, and gasoline runoff). These toxic chemicals tend to accumulate in benthic sediments of urban streams, lakes, and estuaries.

Heavy Metals. Heavy metals are of concern because of their toxic effects on aquatic life and their potential to contaminate drinking water supplies. The heavy metals having the highest concentrations in urban runoff are copper, lead, and zinc, with cadmium a distant fourth. However, when inappropriate connections between sanitary and storm sewers are present, other heavy metals such as arsenic, beryllium, chromium, mercury, nickel, selenium, and thallium can be found. A large fraction of the heavy metals in urban runoff are adsorbed to particulates and, thus, are not readily available for biological uptake and subsequent bioaccumulation. Also, the typical periods of exposure are those of urban runoff events (typically under 8 hours), which are much shorter than the exposure periods used in bioassay tests (typically 24 to 96 hours for toxicity testing). Nonetheless, it is likely that the heavy metals in urban runoff are toxic to aquatic life in certain situations, particularly for the more soluble metals such as copper and zinc. Additionally, resuspension of bottom deposits from high flow events may impact on downstream benthic invertebrates. Compared to risks to aquatic life, human health risks appear to be more remote.

Oil and Grease. Oil and grease contain a wide variety of hydrocarbon compounds, some of which (e.g., polynuclear aromatic hydrocarbons) are known to be toxic to aquatic life at low concentrations. Hydrocarbons are often initially found as a rainbow-colored film or sheen on the water's surface. Other hydrocarbons, especially weathered crankcase oil, appear in solution or in emulsion and have no sheen. However, hydrocarbons have a strong affinity for sediment, and much of the hydrocarbon load eventually adsorbs to particles and settles out. Hydrocarbons tend to accumulate rapidly in the bottom sediments of lakes and estuaries, where they may persist for long periods of time and exert adverse impacts on benthic organisms. The precise impacts of hydrocarbons on the aquatic environment are not well understood. Bioassay data which do exist are largely confined to laboratory exposure tests for specific hydrocarbon compounds. Remarkably few toxicity tests have been performed to examine the effect of urban runoff hydrocarbon loads on aquatic communities under the typical exposure conditions found in urban streams.

Other Pollutants. Other toxic compounds that have been detected in urban runoff include pesticides, herbicides, and synthetic organic compounds. Concentrations of these toxic substances in runoff from residential and commercial areas rarely exceed current water quality criteria. However, it should be noted that there has been relatively little sampling of runoff reported from industrial areas, where toxic compounds might be expected to be more prevalent.

1.2.5 Bacteria

Bacterial levels in undiluted urban runoff usually will exceed federal public health standards for water contact recreation and shellfish harvesting. Because bacteria multiply faster during warm weather, it is not uncommon to find a twentyfold difference in bacterial levels between summer and winter. The substantial seasonal differences often found do not correspond with comparable variations in urban activities. This suggests that in addition to temperature effects, many sources of coliform unrelated to those traditionally associated with human health risk (e.g., animal excrement, illicit connections, leaking sanitary collection systems) may be significant. Thus, despite the high numbers of coliforms found in urban runoff, in the absence of contamination from sanitary sewage, the health implications are unclear. The current literature suggests that fecal coliform may not be consistently reliable in identifying human health risks from urban runoff pollution.[6]

Studies conducted by the National Oceanic and Atmospheric Administration indicate that urban runoff is a major pollutant source which adversely affects shellfish growing waters.[7] The NOAA studies identified urban runoff as affecting over 578,000 acres of shellfish growing waters on the East Coast (39% of the harvest-limited area); 2,000,000 acres of shellfish growing waters in the Gulf of Mexico (59% of the harvest-limited area); and 130,000 acres of shellfish growing waters on the West Coast (52% of harvest-limited areas).

Although nearly every urban and suburban land use can export bacteria at levels which will violate health standards, older and more intensively developed urban areas typically produce the greatest export. The problem is especially significant in urban areas that experience combined or sanitary sewer overflows that export bacteria derived from human wastes.

---

[6] Moffa, P.E. (ed.) 1990. Control and Treatment of Combined Sewer Overflows. Van Nostrand Reinhold, New York, NY.

[7] National Oceanic and Atmospheric Agency. 1990. The Quality of Shellfish Growing Waters on the West Coast of the United States; NOAA. 1989. The Quality of Shellfish Growing Waters on the East Coast of the United States; NOAA. 1988. The Quality of Shellfish Growing Waters in the Gulf of Mexico.

1.2.6 Floatables

Floatable debris in storm water runoff commonly includes plastic and paper products, garden refuse, tires, and metal and glass containers. These pollutants degrade the aesthetic quality of both receiving waters and river banks and shorelines. Vegetation and wildlife may also be impacted. Floating debris may impair restoration efforts by hindering the establishment of emergent vegetation.[8] Fish and aquatic wildlife mortality may also be attributed to debris, due to either ingestion or entanglement in the slowly decomposing materials.

1.3 STORM WATER CONTROL PRACTICES

This section examines general storm water control practices and presents examples of successful storm water programs. A more specific and detailed discussion of various site-specific controls is contained in Chapters 7.

A basic principle of storm water controls for urban development is that it is much more costeffective and institutionally feasible to develop controls for new development than it is to retrofit old development. Structural practices for storm water pollution control require not only capital, but operation and maintenance costs, and are often constrained by spatial and financial limitations in core urban areas. In addition, some structural controls can destroy the resource it is designed to protect due to disruption of the hydrologic cycle. In contrast, nonstructural practices may be included in the development process, for which municipalities usually have preexisting permitting programs. Combining both types of controls into an integrated storm water management program can result in effective water quality protection at minimal cost. Before presenting representative case studies, the practices for control of storm water are first defined.

---

[8] U.S. Army Corps of Engineers. 1990. Anacostia River Basin Reconnaissance Study. U.S. Army Corps of Engineers, Baltimore District, Maryland.

1.3.1 Control Practices

Nonstructural Practices. Nonstructural practices are those not requiring construction or maintenance. These differ from structural practices in that they are preventative in nature, and have the potential to be more cost effective especially if implemented early in the site planning stages. They include such practices as: regional comprehensive storm water management programs; planning future development to minimize storm water runoff; limiting the amount of impervious surface in new and retrofitted development; instituting fertilizer and pesticide management programs; requiring setbacks from surface water and wetlands to protect their environmental integrity; siting infrastructure so as not to encourage development in environmentally sensitive areas that are critical to maintaining water quality; requiring the use of best management practices through land development regulations and site plan approval; and inspecting storm water management systems and erosion control structures to ensure they are functioning properly.

Nonstructural practices for controlling storm water pollution have typically centered around preventing land use disruptions on areas critical to maintaining water quality and reducing the source of pollution.

Examples of Nonstructural Storm Water Practices:

* Zoning Ordinances
* Subdivision Regulations
* Capital Improvement Plans
* Site Plan Reviews
* Planned Unit Development Reviews
* Restrictive Covenants
* Environmental Impact Assessments
* Public Education Programs
* Growth Management
* Buffers and Setbacks
* Environmental Permitting
* Pollution Prevention for All Sources
* Spill Control Programs
* Road Maintenance Programs

Structural Practices. Structural management practices are defined as those designed and constructed to mitigate the adverse impact of storm water runoff. The selection and use of individual practices has typically been based on land use activities, existing structures, hydrology and climate, soil type, and other site-specific conditions.

In addition to installation costs, structural practices usually require continuing operation and maintenance efforts. The lack of adequate maintenance and upkeep may dramatically reduce their effectiveness in removing pollutants from storm water runoff.

Separate storm sewers may also receive materials other than storm water (e.g., illicit connections from industrial and commercial facilities). Controlling these sources may involve structural practices such as conventional wastewater treatment units, or repairing/retrofitting connections to the storm sewer system.

Examples of Structural Storm Water Practices:

Developing Areas

* Extended Detention Ponds
* Storm water Wetlands
* Multiple Pond Systems
* Infiltration Trenches
* Infiltration Basins
* Filter Strips

Core Urban Areas

* Illicit Connection Controls
* Porous Pavements
* Storm water Detention/Wetland Retrofits
* Sand Filters

Integrated Management Programs. The storm water management practices presented above may be used in conjunction with one another, taking an integrated approach to minimizing storm water impacts. Structural practices could be targeted at areas already built up, while developing areas utilize a more nonstructural approach.

Strategies have been shown to be successful when targeted to land disturbance, not necessarily land use, and should reflect land use/activity changes. Guiding development to areas capable of sustaining growth without excessive impacts to the natural environment, and encouraging the implementation of storm water practices as development proceeds, can minimize the need for future storm water control efforts. Many local governments have adopted integrated storm water management programs to regulate development activities within their jurisdictions. Several states, including Oregon, New Jersey, Delaware, and Florida, have adopted comprehensive plans involving guidance of future growth and avoidance of water quality and quantity impacts associated with uncontrolled development.

1.3.2 Land Disturbance/Activity

The extent of storm water pollution problems is dependent upon the land disturbance/activity which in sum is a function of the stage of the urbanization process. The range of storm water management options applied has been based to a certain extent on the stage of land development, each stage representing a unique set of challenges and opportunities.

Undeveloped Areas. Undeveloped areas consist of relatively unurbanized land with low population densities. Although the land use is primarily rural, the proximity and location of these lands presents the potential for eventual development into urban and suburban settings. Storm water runoff from these areas currently results primarily from agricultural, forestry, and resource extraction activities.

In undeveloped areas, storm water management programs may be integrated with local planning and regulatory programs at an early stage of development. These management programs, relying primarily on nonstructural practices, have been aimed at minimizing future degradation of water quality.

Depending on the expected degree and rate of development, integrated management programs have been tailored to address pollution generating activities associated with the various phases of urbanization. Many local governments, aware of the consequences of uncontrolled urbanization, have adopted planning programs in which storm water pollution considerations were major decision criteria.

Lake Travis, Texas Case Study - Phosphorous. The Highland Lakes, a chain of reservoirs in Central Texas, provide hydroelectric power, flood control, and recreational opportunities. Sedimentation, eutrophication, and toxic contamination problems caused by storm water runoff to these lakes prompted the Lower Colorado River Association (LCRA) to develop the Water Quality Leadership Policy (WQLP) in 1988.[9] Analysis of the monitoring data of the Highland Lakes showed that over 90% of the pollution was from nonpoint sources. LCRA estimates that the nonpoint source loads to the Lake Travis basin could increase by 200 to 600% in the future, largely due to the conversion of rangeland into urban and suburban development. The WQLP initiated public education efforts along with a regulatory program to control nonpoint source pollution in its ten-county district. The Lake Travis Nonpoint Source Pollution Control Ordinance was adopted by the LCRA Board of Directors in December 1989 and went into effect on February 1, 1990. The ordinance targets new urban and suburban development in the 250-square mile area of western Travis County.

The Lake Travis Ordinance establishes a set of performance standards that require new developments to remove a specified amount of the annual NPS pollution load, depending on the site's proximity to the shoreline and the slope of the property. Sites within 500 feet of the lake and/or those on steep slopes require a higher level of runoff treatment than those inland or on flatter ground. In general, the more intensively a site is developed, the more pollution must be removed. This approach recognizes the link between land use and NPS pollution; however, there are no land use control stipulations in the ordinance.

The Ordinance requires all land owners proposing to develop land within the Lake Travis watershed to submit an application for review and plans on how the surface runoff from the site will be treated. The plan must include temporary erosion and

[9] Hartigan, P. and K. Wilwerding. 1991. The Clean Colorado Project and Urban Nonpoint Source Pollution Control: The LCRA Program. In Seminary Publication of the Nonpoint Source Watershed Workshop, Sept. 1, 1991, pp. 167-171. EPA/625/4-91/027.

sediment control plans including a restoration program for all disturbed areas, description of the design and location of structural practices used to meet the performance standards, and establishment of a maintenance organization to ensure that the structural practices are adequately maintained. An ongoing program conducted jointly by LCRA, the U.S. EPA Region 6, and the U.S. Geological Survey will evaluate the Ordinance on a periodic basis to determine whether the standards are adequate to protect the water quality of Lake Travis and whether or not the structural best management practices are adequate to meet the standards of on-site pollutant removal.

Developing Urban Areas. Developing urban areas are those lands currently experiencing new development or redevelopment resulting in significant changes to the landscape. These areas include suburban cities and urban fringes located adjacent to urban areas.

Many of these areas are experiencing a dramatic rise in population compared to well-established urban areas. For instance, between the 1970 and the 1980 period, the population of urbanized areas increased 30 times more (an increase of 18.9 million) than core cities with a population exceeding 100,000. One activity affecting storm water quality in developing areas is land disturbance in and around construction sites where exposed soils results in increased sedimentation, erosion, and nutrient transport. Land disturbance in these areas may result in severe storm water pollution if adequate control programs are not incorporated into the development process. Yet these areas, where the majority of new growth and land disturbance and conversion are occurring, are not necessarily covered under the Phase I National Pollutant Discharge Elimination System (NPDES) storm water program. Moreover, an important characteristic of these areas resulting from land conversion consists of permanent changes to land use patterns. These changes induce a significant disruption to the hydrologic cycle and modification of runoff water quality (modification of natural vegetation and infiltration rates, increase storm peak discharges and transport capacities, and decrease of low flow especially during prolonged dry periods).

Storm water management efforts have been directed primarily toward controlling construction site storm water runoff. Storm water pollution control efforts have been minimized by retaining and treating storm water onsite rather than by addressing impacts occurring offsite. To address impacts originating from developing areas, several state and local agencies have developed comprehensive programs incorporating both structural and nonstructural practices.

Dillon Reservoir, Colorado Case Study - Eutrophication / Urban Growth. Dillon Reservoir is a large (2970 acre) impoundment of the Blue River in Colorado. The reservoir supplies drinking water to the Denver metropolitan area and is used for recreation, fish habitat, and agriculture. Water quality degradation, primarily eutrophication and sedimentation, has occurred since the Reservoir's construction over 20 years ago. At higher elevations the watershed is primarily undeveloped, while the lower elevations contain three major municipalities, four major treatment plants, housing developments, and a large molybdenum mine. A study completed in 1983, supported by the EPA through the Clean Lakes Program of the Clean Water Act, identified phosphorus as the primary contributor to the Reservoir's eutrophication. According to the Clean Lakes study, which evaluated 1982 Reservoir phosphorus levels, human activities were found to account for about ½ of the total phosphorus load, and of this, between ½ and 2/3 was attributed to urban nonpoint sources including runoff from parking lots, golf courses, and construction sites, as well as leakage from septic tanks.

The Northwest Colorado Council of Governments, cooperating with local and state agencies, developed a strategy for phosphorus control by holding the phosphorus loadings to existing (1982) waste load allocation levels (4609 kg/yr) and allowing no further water quality degradation.[10] To meet the phosphorus loading while allowing for future growth in the basin, nonpoint source controls were encouraged by allowing

[10] U.S. EPA. 1991. Nutrient Trading in the Dillon Reservoir. Prepared by Apogee Research, Inc. for the U.S. Environmental Protection Agency, Office of Water.

point/nonpoint source trading at a ratio of 2:1. For every 2 kg of nonpoint source phosphorus removed, 1 kg of phosphorus credit is added to the point source limit. This system provided for maintenance of existing phosphorus levels in the Reservoir, while accounting for increased loading from growth. In addition to providing environmental benefits, this program provides economic benefits and incentives for nonpoint source reduction. The 2:1 ratio provides a margin of safety in the source trading program, and if further phosphorus reduction in the Reservoir is desired, the ratio can be adjusted.

To date, only one point/nonpoint source trade has been completed. The low volume of trade has caused the program to operate differently than was first envisioned. The low volume of trade occurred because the publicly owned treatment works (POTWs) were able to achieve some of the highest phosphorus removal efficiencies in the nation using expensive advanced treatment technology. Consequently, point/nonpoint source trading has played only a minor role in the overall basin-wide phosphorus mitigation strategy. The major constraint to future development is a limit on nonpoint source phosphorus loading.

Although total phosphorus loading was only 5559 pounds in 1989, i.e., 54% of the total phosphorus allowed, additional reductions in phosphorus loadings must result from reductions in nonpoint sources. Nonpoint/nonpoint trades are beginning to be used to offset nonpoint sources from new development through the control of existing nonpoint sources.

Core Urban Areas. Existing urban areas, with typical population densities greater than 100,000, are communities with limited potential for new development. These existing, incorporated urban areas with typical populations greater than 100,000 are currently subject to Phase I NPDES storm water permit requirements. The Phase I NPDES storm water program also applies to some highly populated counties.

Original storm water systems in established urban areas were typically constructed for flood control purposes. Water quality programs probably did not address storm water quality concerns, and runoff is typically directed to surface water.

These urban areas are characterized by high percentages of impervious surfaces, which contribute to increased storm water discharges and pollutant transport capacities. Reduction of infiltration rates and groundwater recharge also results in lower baseflows and higher pollutant concentrations in receiving waters, especially during prolonged dry periods. Other storm water impacts include increases in temperature and concentrations of toxic chemicals, nutrients, heavy metals, oil, grease, and pesticides.

Core urban areas may also experience storm water pollution problems resulting from illicit connections, leaking sanitary sewage systems, or groundwater infiltration. Illicit connections can often be traced to the initial development of the storm sewer system, or arise during redevelopment where storm sewers are either mistaken for sanitary sewers or intentionally used for wastewater conveyance. In industrial facilities, floor drains or other discharge points which are connected to the separate storm sewer system may receive spills, rinse waters, or process wastewaters that should be sent to a treatment plant. Illicit connection programs are directed at identifying such problems for corrective action.

Storm water control practices in established urban areas have typically included retrofits of detention ponds and controls on combined sewer systems as well as the construction of wetlands and sand filters. As development proceeds, the range of available nonstructural options decreases.

Kelsey / Bear Creeks, Washington Case Study - Habitat Alteration / Sediments. Kelsey Creek is located in a highly urbanized suburb of Seattle, Washington. The watershed is 90% developed with residential units and commercial/light industry uses.[11] Kelsey Creek, a natural water channel, was developed to convey storm water from the City of Bellevue to Lake Washington, a major water body in the Puget Sound area. The creek serves as a recreational resource and has a productive, but limited, salmon fishery. Increased peak flows from urban development dramatically altered the stream channel, causing severe streambed erosion. The number and diversity of aquatic organisms declined as bed scouring and the resultant deposition

[11] Pitt, R. and P. Bissonnettee. 1984. Bellevue Urban Runoff Program: Summary Report.

of suspended sediment destroyed stream habitat. Reduced dissolved oxygen in the sediments depressed salmon embryo survival. The fish population adapted to the degrading environment by shifting species composition from coho salmon to less sensitive cutthroat trout.

Urban storm water was monitored for six metals and suspended solids. Dry weight concentrations of pollutants from various nonpoint sources, including atmospheric deposition, street dirt accumulation, and catchbasin and detention basin sediment, were measured. Impervious sources (streets, sidewalks, driveways, parking lots, and rooftops) were found to contribute more than 60% of the total runoff flow when precipitation exceeded 0.1 inch. Street surfaces contributed 25% of the total flow in the monitored sites. Most of the total solids in urban runoff originated from residential yards. Nutrients were primarily detected in street dirt samples, originating from vehicle emissions. Only a small fraction of the total particulate loadings on the impervious surfaces were removed by rain (15%). Large particles were not effectively removed, while about one half of the smallest particles (less than 50 microns) were washed off during rains. These small particles were not very abundant, but contained high heavy metal and nutrient concentrations. Most of the settled particulate material in the storm drainage inlets and sewerage pipes was not removed during observed storms.

Several control practices were instituted as pilot projects to determine their effectiveness: street sweeping, catch basin cleaning, and detention ponds. Intensive street sweeping (three times per week) resulted in rapid and significant decreases in street surface suspended solids loadings, from 110 g/curbmeter down to 55 g/curbmeter. This 50% reduction in suspended solids loadings resulted in a maximum 10% decrease in metal loadings. The median particle size also decreased significantly with intensive street cleaning. A regenerative air street cleaner showed substantially higher performance in removing the finer street particles. It appears that conventional street sweeping removes the larger particles and rain removes the smaller particles; however, street sweeping did not reduce loadings of toxic compounds by more than 10%. Cleaning of storm drainage inlets and catch basin sumps twice per year reduced the lead and total solids runoff concentrations by between 10 to 25%. COD, nutrients, and zinc were reduced by between 5 and 10%. After an initial cleaning, it appeared that almost a full year was required for sediment to reach a "stable volume" in the storm drain inlet structures. Only 60% of the total available sump volumes in the inlet structures and catch basins were used for detention of particulates at the "stable volume." Small detention basins (detention times of 30 minutes or less) did not have any significant effect on urban runoff quality, but did reduce peak flow rates by up to 60%. Detention basins should be carefully located so increased flow rates do not disturb critical habitat areas. The final recommendation states that if intensive street sweeping was implemented along with semiannual catch basin sediment removal, urban runoff discharges for most pollutants would be reduced by as much as 25%. Though these reductions are small, they may be important in reducing the accumulation of contaminated sediments in smaller creek systems.

1.3.3 Illicit Connection Controls

Recent studies in Michigan have recognized that development occurring while undersized wastewater treatment plants are operating can create widespread illicit connection problems. For example, the Huron River Pollution Abatement Program in Wastenaw County, Michigan, inspected 660 businesses, homes, and other buildings discharging storm water to the Allen Creek drain. Of the buildings inspected, 14% were found to have improper storm drain connections, with the highest percentage (60%) for automobile-related businesses. Although some of the problems discovered in this study resulted from improper plumbing or illegal connections, the majority were approved connections at the time they were built. Efforts are underway to correct those illicit connections identified during the inspection program.[12]

1.3.4 Combined Sewer Pollution Control

Structural practices for controlling combined sewer overflow pollution usually parallel conventional wastewater treatment practices. They are designed to handle intermittent and random

---

[12] 40 C.F.R. Parts 122, 123, and 124.

flows which vary in magnitude and quality. These practices include in-line storage, off-line storage, storage sedimentation, swirl concentrators, screens, dissolved air floatation, high rate filtration, treatment lagoons, contact stabilization, rotating biological contactors, and high rate trickling filters. They all differ in costs, efficiencies, suitability, and operation and maintenance.

In-line storage control. In-line storage control is a low capital cost method that uses existing facilities. It is easily integrated with dry weather collection, treatment, and disposal activities, and is adaptable to future expansion. Metropolitan Seattle implemented an on-line storage control system to mitigate Combined Sewer Overflows (CSO) impacts to receiving waters.[13] The system performance was evaluated in terms of the reduction of overflow volume, overflow frequency, and annual reduction of pollutant loads.

The system was estimated to reduce overflow volume by 74%. The resulting reductions in pollution loads were estimated to be 493,000 kg/year of suspended solids, and 136,000 kg/year of BOD_5. The total in-line storage control costs for this system are summarized as follows:

Regulators	$9,762,000
Computer facilities	$5,717,000
Engineering	$ 924,000
O & M cost (per year)	$ 440,000

Saginaw, Michigan uses a combination of storage and treatment processes to take advantage of the capabilities of existing systems.[14] This integrated approach reduces the volume of overflows to receiving waters and treats overflows that do occur to near primary treatment levels. This integrated system at the Hancock Street facilities includes:

[13]Finnemore, E. John. 1982.Stormwater Pollution: Structural Measures. Journal of Environmental Engineering Division, Proceeding of the American Society of Civil Engineers, Vol. 108, No. EE4.
[14]Id.

1. In-line storage;
2. Using existing interceptor capacity controlled by modified regulator stations;
3. A flood protection pumping station, an off-line storage treatment basin capable of treating and disinfecting all overflows; and
4. A capability to treat all flows retained in storage at the local dry weather treatment plant.

Together, this system supports a city-wide plan to eliminate uncontrolled combined sewer discharge to the Saginaw River.

The performance of the Hancock Street storage/treatment facility was characterized using data from the 1978 summer monitoring program. Eleven storms occurred during the monitoring period and an estimated 52,000,000 gallons were pumped to the storage/treatment basin. Of this volume, 40% overflowed, after treatment, during three storms. The overflow frequency was reduced by 73%, and the effectiveness of the basin in treating the overflows ranged between 35 and 75% reduction in concentrations of eight different pollutants.

The total construction cost of the Hancock Street CSO control facilities was $7,280,000, including modifications to the in-line storage system (regulators) and the storage/treatment facilities. Estimated annual operating and maintenance cost for the storage/treatment system is about $50,000/yr.

1.3.5 Detention Ponds and Retention Basins

The ability of wet detention ponds and retention basins to remove pollutants from storm water has been extensively studied. Numerous case studies have shown that reductions in suspended sediments, nutrients, and heavy metals are possible through the use of either retrofitted storm water basins or detention ponds designed specifically for water quality improvements. The following case study reported variable reductions in total suspended solids, metals, and nutrients.

A three-pond storm water retention system receiving storm water from a highway interchange in Orlando, Florida, was investigated.[15] The three ponds were interconnected by way of a large culvert to allow overflow to other ponds when storm runoff exceeds the design level. The first receiving pond has an approximate surface area of 1.3 ha (3.2 acres), an average depth of 1.5 m (4.9 ft), and a total drainage area of 10.8 ha (26.7 acres). The ponds maintained a large standing crop of filamentous algae virtually yearround.

Field investigations conducted during 1982-1984 were designed to assess (1) the quantity of pollutants entering the pond, (2) the average water quality parameters in the basin water, (3) the accumulation of nutrients and heavy metals in the sediment of the pond, and (4) the leaching of heavy metals to the groundwater beneath the retention ponds.

The removal efficiencies of particulate metals were found to range from 77% for copper to over 95% for lead and zinc, while the removal of the dissolved fraction was only about 50% for lead and copper and 88% for zinc. The removal of particulate phosphorus and organic nitrogen was, on the other hand, poor and did not exceed 12%. The removal of the dissolved fraction of nutrient loadings ranged from 81.6% for ammonia to 90% for phosphorus.

Accumulation of phosphorus in the bottom sediment of the pond was evaluated at 99% of the total input during a 7-year period. However, 85-90% removal of the total nitrogen load was attributed to nitrification-denitrification processes. The removal of particulate heavy metals from the pond water was also attributed to settling and accumulation in the bottom sediment within a relatively short distance from the storm water inlet.

One common concern of using retention basins is the potential for groundwater contamination. However, monthly groundwater samples revealed no conclusive evidence of heavy metals migration from the pond.

---

[15] Youssef, Y.A., M.P. Wanielista, and H.H. Harper. 1986. Design and Effectiveness of Urban Retention Basins. In: B. Urbonas and L.A. Roesner (eds.), Urban Runoff Quality - Impact and Quality Enhancement Technology.

1.3.6 Sand Filters

The State of Delaware developed a design approach for implementation of storm water management water quality control practices targeting urban areas having little or no pervious area.[16] The controls were adopted under the urban storm water runoff component of the NPDES program. The State of Delaware enacted an urban storm water control program focusing on new construction. After July 1, 1991, all new development activities require reviews for water quality impacts prior to their approval. The program requires that predevelopment peak discharges are not exceeded and water quality is protected by vegetative and structural control strategies designed to remove 80% of the inflow's suspended solids. Shallow storm water management ponds that encourage plant growth are preferred because nutrients are a significant source of water quality degradation in Delaware. Often spatial or other engineering considerations limit the use of ponds in urban areas. In heavily urbanized areas, sand filters may be appropriate for both new development and retrofits as they do not limit land usage. The sand filters will be designed to treat runoff from frequent storm events, i.e., the first inch of rainfall. The design procedures for the sand filters are based on equations developed by the City of Austin, as well as structural design previously used for a project in the State of Maryland.

Filter efficiency is related to the distribution of pollutants in the various particle size classes. Most urban particulate matter is of a coarser size fraction; however, most of the other pollutants (except for metals, which tend to be more evenly distributed across the size classes) are associated with the smaller particle sizes. For example, approximately 6% of urban particulate matter is in the silt and clay soil size range; but silts and clays contain more than 50% of the phosphorus. The sedimentation chamber of the filtration system is designed to remove the sand and gravel components and the sand filter is designed to remove the finer silt and clay particles. The detention chamber provides a detention time sufficient to remove sand and coarse particles based on their calculated settling velocities. The filter drains the

[16] Shaver, E. 1991. Sand Filter Design for Water Quality Treatment. Presented at 1991 ASCE Stormwater Conference in Crested Butte, Colorado.

first inch of runoff within a 24-hour period by assuming an average filtration rate of 0.04 gal/min/ft for the sand. The design drainage area, typically a parking lot, must be less than 5 acres. Predicted removal efficiencies are 70% removal of total suspended solids, 33% removal of total phosphorus, 21% removal of total nitrogen, and 45% removal of the metals lead and zinc. Actual removals contained in the Austin report are somewhat higher than expected.

Chapter 2

STORM WATER DISCHARGE REGULATION

2.1 CLEAN WATER ACT NPDES PERMIT PROGRAM

Due to the complexity of the problem, government regulation of storm water discharges was debated for many years. The Clean Water Act (CWA)[17] - specifically, its National Pollutant Discharge Elimination System (NPDES)[18] - was the

[17] Clean Water Act (CWA), 33 U.S.C. 1251-1386.

[18] CWA 402, 33 U.S.C. 1342. In 1972, the Clean Water Act was amended to provide that the discharge of any pollutant to waters of the United States from any point source is unlawful, except if the discharge is in compliance with a National Pollutant Discharge Elimination System (NPDES) permit.

obvious means for regulating storm water discharges. The CWA requires that point source discharges must be authorized by a NPDES permit. NPDES permits are generally issued by states with EPA-approved NPDES programs. Most states have approved programs. For those states without NPDES permit authority, EPA Regional offices for those states oversee issuance of NPDES permits. Due to limited resources, efforts to control point source pollution through the NPDES permit program have primarily focused on controlling pollution discharges from publicly owned treatment works (POTWs) and industrial process wastewaters. Under CWA Section 402(p), the NPDES permit program also addresses storm water discharges.

Under the NPDES system, permits are issued to different types of facilities that discharge pollutants into U.S. waters. A NPDES permit is essentially a license that allows a facility to discharge contaminated water. On the NPDES permit application, the facility provides information about the type of facility and type of discharge being requested. If the facility's application is approved, the permitting authority will issue a permit that contains various conditions related to the facility's pollutant discharges. The permit will generally contain specific limitations on contamination levels or specific actions that the facility must take, such as sampling or inspections.

There are two different types of permits that are issued under the CWA's NPDES program: individual permits and general permits. An individual permit is issued for one particular facility. A permit writer reviews the facility's NPDES application and may make a site visit to better understand the facility's operation, and then tailors the individual permit to address specific conditions at the facility. Although the permit is individualized, a large portion of the permit will, however, consist of standard conditions. These conditions are included in all permits issued by the permitting authority. Standard conditions often include the following:

* Duty to provide information
* Signatory requirements
* Inspection and entry
* Penalties for falsification of records
* Definitions

* Notification requirements
* Nontransferability
* Use of proper operation and maintenance
* Halt production for noncompliance
* Duty to reapply

Facility-specific conditions usually include monitoring requirements which require that the facility take samples of the discharges into U.S. waters and analyze the contaminant levels of specified pollutants. The most commonly requested pollutants to be analyzed are pH, total suspended solids, and oil and grease. Sampling of additional pollutants may be required if they are suspected to be in the facility's discharges.

For a number of reasons, EPA and authorized NPDES States failed to issue NPDES permits for the majority of point source discharges of storm water. Recognizing this, Congress added Section 402(p) to the CWA in 1987 to establish a comprehensive framework for addressing storm water discharges under the NPDES program.[19]

2.2 EPA's STORM WATER PROGRAM

CWA Section 402(p) establishes a comprehensive two-phased approach for EPA to address storm water discharges. Under Phase I, NPDES permits are required to be issued for municipal separate storm sewers (MS4s) serving large or medium-sized populations (greater than 250,000 or 100,000 people, respectively) and for storm water discharges associated with industrial activity. Permits are also to be issued, on a case-by-case basis, if the EPA or authorized state determines that a storm water discharge contributes to the violation of a water quality standard or is a significant contributor of pollutants to U.S. waters.

Under Phase II, EPA is to prepare two reports to Congress that assess remaining storm water discharges not covered under the Phase I regulations. The Phase II regulations were to have

[19] 33 U.S.C. 1342(p). Congress amended the Clean Water Act with the Water Quality Act of 1987, Pub. L. No. 100-4, 101 Stat. 7 (codified as amended in scattered Sections of 33 U.S.C. 1251-1386).

been issued by EPA not later than October 1, 1992, however, because it took so long to finalize the Phase I program, EPA has not yet been able to complete and issue appropriate regulations for the Phase II program. However, proposed regulations were under consideration at the time of this writing.[20] Table 2.1 summarizes the most important EPA rulemaking activities concerning the storm water program.

Table 2.1 EPA Rulemakings for Storm Water Discharges

Date	Action	Title	Citation
4/7/95	Direct Final Rule	Amendment to Requirements for National Pollutant Discharge Elimination System (NPDES) Permits for Storm Water Discharges Under Section 402(p)(6) of the Clean Water Act	60 Fed. Reg. 17950
11/19/93	Notice	Fact Sheet For the Multi-Sector Storm Water General Permit	58 Fed. Reg. 61146
12/18/92	Final Rule	NPDES; Storm Water Discharges; Permit Issuance and Permit Compliance Deadlines for Phase I Storm Water Discharges	57 Fed. Reg. 60444
9/30/92	Correction	Final NPDES General Permits for Storm Water Discharges From Construction Sites	57 Fed. Reg. 45112

[20] "Amendment to Requirements for National Pollutant Discharge Elimination System (NPDES) Permits for Storm Water Discharges Under Section 402(p)(6) of the Clean Water Act," 60 Fed. Reg. 17950 (Apr. 7, 1995); "Request for Comment on Alternative Approaches for Phase II Storm Water Program," 57 Fed. Reg. 41344 (Sept. 9, 1992);

Date	Action	Title	Citation
9/25/92	Notice	Final NPDES General Permits for Storm Water Discharges From Construction Sites	57 Fed. Reg. 44412
9/25/92	Notice	Final NPDES General Permits for Storm Water Discharges Associated With Industrial Activity	57 Fed. Reg. 44438
9/9/92	Proposed Rule	NPDES, Request for Comment on Alternative Approaches for Phase II Storm Water Program	57 Fed. Reg. 41344
9/9/92	Notice	Final NPDES General Permits for Storm Water Discharges from Construction Sites	57 Fed. Reg. 41176
9/9/92	Notice	Final NPDES General Permits for Storm Water Discharges Associated With Industrial Activity	57 Fed. Reg. 41236
4/2/92	Final Rule	NPDES Application Deadlines, General Permit Requirements and Reporting Requirements for Storm Water Discharges Associated With Industrial Activity	57 Fed. Reg. 11394
11/16/90	Final Rule	NPDES Permit Application Regulations for Storm Water Discharges	55 Fed. Reg. 47990

2.2.1 Phase I Permit Regulations

Five classes of storm water discharges are specifically listed under Section 402(p)(2) which are covered under "Phase I" of the storm water program:

1. A discharge with respect to which a permit has been issued prior to February 4, 1987;
2. A discharge associated with industrial activity;
3. A discharge from a municipal separate storm sewer system serving a population of 250,000 or more;
4. A discharge from a municipal separate storm sewer system serving a population of 100,000 or more, but less than 250,000; or
5. A discharge for which the EPA Administrator or the NPDES-authorized State, as the case may be, determines that the storm water discharge contributes to a violation of a water quality standard or is a significant contributor of pollutants to U.S. waters.

Section 402(p)(3) confirms that, like all other point source discharges under the CWA, discharges of storm water associated with industrial activity must meet all applicable provisions of CWA Sections 402 and 301, including technology-based requirements and any necessary water quality-based requirements. Permits for discharges from municipal separate storm sewer systems may be issued on a system- or jurisdiction-wide basis and must meet a new statutory standard requiring controls to reduce pollutant discharges to the maximum extent practicable (MEP).

EPA promulgated the Phase I permit application regulations for storm water discharges on November 16, 1990.[21] These regulations established the scope of the Phase I storm water program by defining two major classes of storm water discharges identified under Section 402(p)(2)(B), (C), and (D) of the CWA: storm water discharges associated with industrial activity; and discharges from municipal separate storm sewer systems (MS4s) serving a population of 100,000 or more. In addition, the 1990 regulations established permit application

[21] "National Pollution Discharge Elimination System Permit Application Regulations for Storm Water Discharges," 55 Fed. Reg. 47990 (Nov. 16, 1990).

requirements, including deadlines for these two classes of discharges.[22]

The regulations defined a municipal separate storm sewer system serving a population of 100,000 or more to include municipal separate storm sewers within the boundaries of 173 incorporated cities, and within unincorporated portions of 47 counties that were identified as having populations of 100,000 or more in unincorporated, urbanized portions of the county.[23]

In addition, the regulations allowed for additional municipal separate storm sewers to be designated by the Director of the NPDES program as being part of a large or medium MS4. The November 16, 1990 regulations establish comprehensive two part permit applications for discharges from large or medium MS4s. The permit application requirements for large and medium MS4s, among other things, require municipal applicants to propose municipal storm water management programs to control pollutants to the maximum extent practicable (MEP) and to effectively prohibit nonstorm water discharges to the MS4.[24]

The 1990 regulations also defined the term "storm water discharges associated with industrial activity" to include 11 categories of industrial facilities.[25] These regulations present three permit application options for storm water discharges associated with industrial activity.[26] The first option was to submit an individual application consisting of Forms 1 and 2F. The second option was to become a participant in a group application.[27] The third option was to file a Notice of Intent (NOI) to be covered under a general permit in accordance with the requirements of an issued general permit.

The Phase I storm water program takes two very different approaches to defining the roles of EPA and authorized NPDES States in controlling pollutants in storm water discharges. With

---

[22]See Table 2.3.

[23]See Appendices F, G, H, and I to 40 C.F.R. part 122.

[24]See 40 C.F.R. 122.26(d)(2)(iv).

[25]See 40 C.F.R. 122.26(b)(14).

[26]See Chapter 3 for a detailed discussion of each of the different permit options.

[27]EPA later proposed an additional type of general permit referred to as a multisector permit. See "Fact Sheet For the Multi-Sector Storm Water General Permit," 58 Fed. Reg. 61146 (Nov. 19, 1993).

respect to permits for large and medium MS4s, the efforts of the NPDES permitting authority (EPA or an authorized NPDES State) are directed to ensuring that municipalities develop and implement storm water management programs to control pollutants to the MEP.

2.2.2 Municipal Activities

Municipal programs address the control of pollutants in storm water from all areas within the boundaries of the MS4 that discharge to the system, including privately-owned lands, as well as modifying municipal activities (e.g., road deicing and maintenance, flood control efforts, maintenance of municipal lands, etc.) to address storm water quality concerns. The agency has defined the role of municipalities under this program in a flexible manner that allows local governments to assist in defining priority pollutant sources within the municipality, and to develop and implement appropriate controls for such discharges. With respect to permits for storm water discharges associated with industrial activity, the NPDES permitting authority has a more direct role in regulating facilities.[28]

2.2.3 Industrial Activities

EPA has defined the term "storm water discharge associated with industrial activity" in a comprehensive manner to address over 100,000 facilities. All storm water discharges associated with industrial activity that discharge directly to U.S. waters or through MS4s are required to obtain NPDES permits, including those which discharge through systems located in municipalities with populations of less than 100,000. Discharges of storm water to a combined sewer system or to a POTW are excluded. Facilities with storm water discharges associated with industrial activity include: manufacturing/industrial facilities; hazardous waste treatment, storage, or disposal facilities; landfills; certain sewage treatment plants; recycling facilities; power plants; mining operations; some oil and gas operations; airports; and certain other transportation facilities. Operators of industrial

---

[28] A detailed description of the role of municipalities in addressing industrial storm water sources under this federal/state/municipal partnership is provided at 56 Fed. Reg. 40972 (Aug. 16, 1991).

facilities that are federally, state or municipally owned or operated (with the exception of certain facilities owned or operated by a municipality of less than 100,000 people)[29] that meet the description of the facilities listed in 40 C.F.R. 122.26(b)(14) (i)-(xi) must also submit applications. Table 2.2 summarizes the industrial activities covered by the Phase I storm water program.

Table 2.2 - Industrial Activities Covered Under the Phase I Program

40 C.F.R. 122.26(b)(14) Subsection	
(i)	Facilities subject to storm water effluent limitation guidelines, new source performance standards, or toxic pollutants effluent standards under 40 C.F.R., Subchapter N (except facilities which are exempt under category (xi)).
(ii)	Facilities classified as: SIC 24 (except 2434) -Lumber and wood products. SIC 26 (except 265 and 267) - Paper and allied products. SIC 28 (except 283 and 285) - Chemicals and allied products. SIC 29 - Petroleum and coal products. SIC 311 - Leather tanning and finishing. SIC 32 (except 323) - Stone, clay and glass products. SIC 33 - Primary metal industries. SIC 3441 - Fabricated structural metal. SIC 373 - Ship and boat building and repairing.

[29] In the Intermodal Surface Transportation Efficiency Act of 1991, Congress provided that industrial activities owned or operated by municipalities with a population of less than 100,000 be placed into Phase II of the storm water program with the exception of airports, power plants, and controlled sanitary landfills.

40 C.F.R. 122.26(b)(14) Subsection	
(iii)	Facilities classified as: SIC 10 - Metal mining. SIC 11 - Anthracite mining. SIC 12 - Coal mining. SIC 13 - Oil and gas extraction. SIC 14 - Nonmetallic minerals, except fuels.
(iv)	Hazardous waste treatment, storage, or disposal facilities, including those that are operating under interim status or a permit under Subtitle C of the Resource Conservation and Recovery Act (RCRA).
(v)	Landfills, land application sites, and open dumps that receive or have received any industrial wastes including those that are subject to regulation under RCRA subtitle D.
(vi)	Facilities involved in the recycling of material, including metal scrapyards, battery reclaimers, salvage yards, and automobile junkyards, including but limited to those classified as: SIC 5015 - Motor vehicle parts, used. SIC 5093 - Scrap and waste materials.
(vii)	Steam electric power generating facilities, including coal-handling sites.

40 C.F.R. 122.26(b)(14) Subsection	
(viii)	Transportation facilities covered by the following SIC codes which have vehicle maintenance (including vehicle rehabilitation, mechanical repairs, painting, fueling, and lubrication), equipment cleaning operations, or airport deicing operations, or which are otherwise listed in another category, are included: SIC 40 - Railroad transportation. SIC 41 - Local and suburban transit. SIC 42 (except 4221-25) - Motor freight and warehousing. SIC 43 - U.S. Postal Service. SIC 44 - Water transportation. SIC 45 - Transportation by air. SIC 5171 - Petroleum bulk stations and terminals.
(ix)	Treatment works treating domestic sewage or any other sewage sludge or wastewater treatment device or system, used in the storage, treatment, recycling, and reclamation of municipal or domestic sewage, including lands dedicated to disposal of the sewage sludge that are located within the confines of the facility, with a design flow of 1.0 million gallons per day (MGD) or more, or required to have an approved pretreatment program under 40 C.F.R. Part 403. Not included are farm lands, domestic gardens, or lands used for sludge management where sludge is beneficially reused and which are not physically located in the confines of the facility, or areas that are in compliance with Section 405 of the CWA.

40 C.F.R. 122.26(b)(14) Subsection	
(x)	Construction activity including clearing, grading, and excavation activities except operations that result in the disturbance of less than 5 acres of total land area which are not part of a larger common plan of development or sale.*
(xi)	Facilities under the following SICs (which are not otherwise included in categories (ii)-(x)), including only storm water discharges where material-handling equipment or activities, raw materials, intermediate products, final products, waste materials, byproducts, or industrial machinery are exposed to storm water:* SIC 20 - Food and kindred products. SIC 21 - Tobacco products. SIC 22 - Textile mill products. SIC 23 - Apparel and other textile products. SIC 2434 - Wood kitchen cabinets. SIC 25 - Furniture and fixtures. SIC 265 - Paperboard containers and boxes. SIC 267 - Converted paper and paper board products (except containers and boxes). SIC 27 - Printing and publishing. SIC 283 - Drugs. SIC 285 - Paints, varnishes, lacquer, enamels. SIC 30 - Rubber and misc. plastics products. SIC 31 - (except 311) - Leather and leather products.

---

40 C.F.R.
122.26(b)(14)
Subsection

(xi) .... con't

SIC 323 - Products of purchased glass.

SIC 34 (except 3441) - Fabricated metal products.

SIC 35 - Industrial machinery and equipment, except electrical.

SIC 36 - Electronic and other electric equipment.

SIC 37 (except 373) - Transportation equipment.

SIC 38 - Instruments and related products.

SIC 39 - Miscellaneous manufacturing industries.

SIC 4221 - Farm products warehousing and storage.

SIC 4222 - Refrigerated warehousing and storage.

SIC 4225 - General warehousing and storage.

---

* On June 4, 1992 the United States Court of Appeals for the Ninth Circuit found that EPA's rationale for exempting construction sites of less than five acres and certain uncontaminated storm water discharges from category xi light industrial facilities from Phase I of the storm water program to be invalid. NRDC v. EPA, 966 F.2d 1292 (9th Cir. 1992).

Source: 55 Fed. Reg. 48065 (Nov. 16, 1990).

2.2.4 Legal Challenges

The regulations defined which discharges are "associated with industrial activity" and thus are subject to permitting under Phase I. EPA included facilities in ten different major industrial categories.[30] One of the categories included discharges from construction sites larger than 5 acres.[31] Under the regulations, facilities in these categories were presumed to discharge storm water associated with industrial activity and were required to submit permit applications.

EPA also established an eleventh category of industrial facilities in "light" industries.[32] For these discharges, by contrast, EPA presumed that there would be no storm water discharge associated with industrial activity, and did not require permit applications unless there was actual exposure of industrial pollutants to storm water at the facility.

The 1990 regulations also addressed requirements, including deadlines,[33] for two sets of application procedures for those storm water discharges which EPA classified as associated with industrial activity: Individual permit applications and group applications. In addition, the notice recognized a third set of application procedures for storm water discharges associated with industrial activity: those associated with general permits.[34]

In 1992, the American Mining Congress challenged EPA's designation of abandoned mining sites as a major industrial category under the Phase I permitting regulations. The court held that EPA was justified in its determination that inactive sites should be included "because some mining sites represent a significant source of contaminated storm water runoff."[35] Further, the court found EPA's regulation of inactive mines reasonable because it was limited to "those sites at which storm

---

[30] 40 C.F.R. 122.26(b)(14)(i)-(x).

[31] 40 C.F.R. 122.26(b)(14)(x).

[32] 40 C.F.R. 122.26(b)(14)(xi).

[33] See Section 2.2.7

[34] The different types of permit applications are discussed in detail in Chapter 3.

[35] American Mining Congress v. EPA, 965 F.2d 759 (9th Cir. 1992).

water discharge is likely to have become contaminated through association with industrial activity."[36]

Another legal challenge was brought by the Natural Resources Defense Council (NRDC) in 1992,[37] objecting to EPA's exemption for oil and gas activities. This challenge focused on the disparity of treatment between oil and gas operations and mining operations.[38] The regulations require owners of oil and gas facilities to apply for a permit only if the facility has a discharge of a reportable quantity,[39] whereas mining operators must submit applications whenever storm water comes into contact with overburden or other listed materials. The court upheld the exemption for oil and gas operations, stating that "... the determination of whether storm water is contaminated is within the [EPA] Administrator's discretion."

2.2.5 Light Industries Exemption

In the same case, the NRDC also challenged the EPA's so-called "light industries" exemption from the definition of "associated with industrial activity" if storm water does not come in contact with these industrial activities.[40] "Light industries" include businesses with certain Standard Industrial Classification (SIC) numbers, such as manufacturers of:

---

[36] EPA excluded certain inactive mines that may not be regulated efficiently, including: (1) inactive mines without an identifiable owner or operator; (2) those that have been "reclaimed" under Surface Mining Control and Reclamation Act or other state laws; (3) those with minimal disturbances (such as those that are undisturbed or where mining operations are solely for the purpose of maintaining a claim); and (4) those where storm water does not come in contact with overburden or other materials. 40 C.F.R. 122.26(b)(14)(iii).

[37] Natural Resources Defense Council v. EPA, 966 F.2d 1292 (9th Cir. 1992).

[38] Section 402(l) of the CWA specifically exempts storm water discharges from mining and oil and gas operations if the discharges are "composed entirely of flows . . . which are not contaminated by contact with, or do not come in contact with," any materials or waste products. 33 U.S.C. 1342(l)(2).

[39] EPA relied on the use of reportable quantities (RQs) in CWA 311(b)(4), 33 U.S.C. 1321(b)(4).

[40] 40 C.F.R. 122.26(b)(14).

* Pharmaceuticals
* Paints
* Varnishes
* Lacquers
* Enamels
* Machinery
* Computers
* Electrical equipment
* Transportation equipment
* Glass products
* Fabrics
* Furniture
* Paper board
* Food processors
* Printers
* Jewelry
* Toys
* Tobacco products

These facilities must acquire a permit for storm water runoff only if there are areas "where material-handling equipment or activities, raw materials, intermediate products, final products, waste materials, by-products, or industrial machinery at these facilities are exposed to storm water."

EPA justified this exclusion for light industries because "most of the activity at these types of manufacturers takes place indoors, and that emissions from stacks, use of unhoused manufacturing equipment, outside material storage or disposal, and generation of large amounts of dust and particles will all be minimal." EPA determined that "these industries are more akin or comparable to businesses, such as retail, commercial, or service industries . . . and storm water discharges from these facilities are not 'associated with industrial activity.'"[41]

The Ninth Circuit struck down EPA's exclusion of certain SIC categories from the definition of "associated with industrial activity." The court held that EPA did not provide any facts in the record to support its determination; therefore, it was arbitrary for EPA to include this added criteria in the case of

[41] 55 Fed. Reg. 47,990, at 48,008

"light industries" when other industries are presumed to be discharging contaminated storm water.

2.2.6 Construction Operations

The NRDC also challenged the EPA's exemption for construction sites under five acres. EPA included construction operations disturbing more than five acres of land in the definition of "associated with industrial activity."[42] EPA chose a five-acre threshold because the agency wanted to limit the amount of permit applications and because it felt that "the acreage limit [was] appropriate for identifying sites that are [sic] amount to industrial activity."[43]

NRDC argued that the exemption was unauthorized because EPA did not have the power to exempt any discharges associated with industrial activity. NRDC also argued that the five-acre limit was arbitrary, especially in light of an acknowledgement by EPA that "[e]ven small construction sites may have a significant negative impact on water quality in localized areas."[44] The Ninth Circuit agreed with NRDC and struck down this part of the regulations. The court noted that EPA conceded that construction was an industrial activity, and therefore, EPA could not create exemptions from the permitting requirements. Furthermore, the court ruled that a five-acre limit was arbitrary due to the lack of factual findings by EPA and due to the admission that even small sites may have a significant impact.

2.2.7 Compliance Deadlines for Phase I Permit Applications

CWA Section 402(p)(4) provided that EPA/States were to issue or deny permits by February 4, 1991 for discharges associated with industrial activity or discharges from large municipal separate storm sewer systems, and by February 4, 1993 for discharges from medium municipal separate storm sewer systems. These dates represent one year after the statutory permit application deadline and two years after the deadline for promulgation of permit application regulations. However, EPA

---

[42] 40 C.F.R. 122.26(b)(14)(x).

[43] 55 Fed. Reg. 47,990, at 48,036.

[44] Id. at 48,033.

was unable to promulgate its regulations by the statutory deadline, and therefore established permit application deadlines which are somewhat later than those specified in the statute, i.e., October 1, 1993 for most discharges associated with industrial activity, May 17, 1994 for discharges associated with industrial activity from municipalities with a population less than 250,000 who participated in a group application, November 18, 1993 for discharges from large municipal systems, and May 17, 1994 for discharges from medium municipal systems.

For applications for new discharges, or existing discharges at facilities that fail to submit a complete permit application by the deadline, EPA or the State must issue or deny the permit within one year after the actual receipt of the permit application. The permit application deadlines are summarized in Table 2.3.

Table 2.3 - Storm Water Permit Compliance Deadlines

Type of application	Permit application deadline	Permit issuance deadline
Industrial Permits		
Individual:		
Existing facilities	October 1, 1992	October 1, 1993.
Facilities rejected from a group application	October 1, 1992	October 1, 1993.
New facilities	180 days prior to commencement of industrial activity which may cause a storm water discharge	One year after receipt of complete permit application
New construction facilities	90 days prior to commencement of construction	One year after receipt of complete permit application

Type of application	Permit application deadline		Permit issuance deadline
	Part 1	Part 2	
Group Permits			
All industrial activities except those owned or operated by a municipality with a population of less than 250,000	Sept. 30, 1991	Oct. 1, 1991	Oct. 1, 1993
Industrial activities owned or operated by a municipality with a population of less than 250,000	May 18, 1992	May 17, 1993	May 17, 1994
Municipal Permits			
	Part 1	Part 2	
Large Municipal Systems	Nov. 18, 1991	Nov. 16, 1992	Nov. 16, 1993
Medium Municipal Systems	May 18, 1992	May 17, 1993	May 17, 1994

For all initial storm water permits, the permittee must comply with all permit conditions no later than 3 years after the date of permit issuance.

2.2.8 Phase II Permit Regulations

CWA Sections 402(p)(5) and (6) require EPA to identify storm water discharges not covered under Phase I which should be regulated to protect water quality.[45] As originally adopted,

[45]Section 502(14) of the CWA excludes agricultural storm water runoff from the definition of point source. Section 402(1)(2) prohibits EPA from requiring an NPDES permit for certain "uncontaminated" storm water

Section 402(p)(1) specified that the "moratorium" on permitting of these storm water sources, which EPA refers to as "Phase II" sources, would expire on October 1, 1992. Section 312 of the Water Resource Development Act of 1992, which was signed by the President on October 31, 1992, extended the moratorium to October 1, 1994. On September 9, 1992, EPA issued proposed regulations for Phase II to identify additional categories of storm water activities for further controls.[46] The proposed regulations suggest cost-effective options for addressing risks from storm water sources not currently required to be permitted that could potentially be addressed under Phase II of the storm water program.

Phase II potentially includes all point source discharges of storm water to U.S. waters (including MS4s) that are not regulated under Phase I of the storm water program. Table 2.4 summarizes the facilities potentially subject to the final Phase II permit regulations.

On April 7, 1995, EPA promulgated a direct final rule with changes to its NPDES storm water permit application regulations to establish a sequential application process for all Phase II storm water discharges.[47]

Phase II Application deadlines are in two tiers. The direct final rule provides the NPDES permitting authority with flexibility to target those Phase II dischargers that are contributing to a water quality impairment or are a significant contributor of pollutants for permitting within the next six years. All other Phase II dischargers are required to apply for a permit only after six years, and only if the Phase II regulatory program in place at that time requires such applications.

EPA has also invited the regulated community to assist in the development of additional Phase II rules, which will be finalized by March 1, 1999. These rules will determine the

---

discharges from mining sites and oil and gas operations. EPA cannot regulate these discharges under Section 402(p)(6) of the CWA.

[46] 57 Fed. Reg. 41344 (Sept. 9, 1992).

[47] "Amendment to Requirements for National Pollutant Discharge Elimination System (NPDES) Permits for Storm Water Discharges Under Section 402(p)(6) of the Clean Water Act," 60 Fed. Reg. 17950 (Apr. 7, 1995).

nature and extent of requirements, if any, that will apply to the various types of Phase II facilities. Both the amendments issued under the direct final rule, as well as the development of the comprehensive Phase II program through an inclusionary process, is a response by EPA to the direction of the President on February 21, 1995, regarding regulatory reform. The final rule goes into effect on August 2, 1995.

Table 2.4 Potential Universe of Facilities Covered Under the Phase II Program

Statutory / Regulatory Exemptions:	- Non Point Source Silviculture Activities. - Agricultural Runoff and Irrigation Return Flows. - Uncontaminated discharges from Mining, Oil and Gas Operations.
General categories of sources	- All municipalities with populations less than 100,000. - All industrial activities not regulated under Phase I (including those owned/operated by municipalities under 100,000) (tank farms, "auxiliary facilities"). - Commercial activities with industrial components (gas stations, dry cleaners). - Construction activities involving less than 5 acres. - Large parking lots (shopping malls, stadiums). - Residential property. - Recreational areas (ski areas, golf courses, amusement parks). - Livestock facilities (stables, feedlots not addressed by Phase I regulations. - Greenhouses, nurseries.

2.2.9 Pollution Prevention Plan Requirements

A storm water pollution prevention plan is a mandatory component for each facility covered by a storm water permit. Storm water pollution prevention plans must be prepared in accordance with good engineering practices and in accordance with the factors outlined in 40 C.F.R. 125.3(d) (2) or (3) as appropriate. The plan must identify potential sources of pollution that may reasonably be expected to affect the quality of storm water discharges from the facility. In addition, the plan must describe and ensure the implementation of practices to reduce the pollutants in storm water discharges associated with industrial activity at the facility and to assure compliance with the terms and conditions of the permit.

The pollution prevention plan must consist of several different components. Each of the different storm water permits described in Chapter 3 have specific pollution prevention requirements. Chapters 5-8 provide detailed procedures for developing and implementing best management practices necessary for satisfying the pollution prevention plan requirement of the storm water permits. Appendix C contains a sample pollution prevention plan for an industrial facility.

2.3 NONPOINT SOURCE POLLUTION CONTROL PROGRAMS

A particularly complex and difficult environmental management problem, which is related to storm water discharges, is nonpoint source (NPS) pollution. After achieving marked success in regulating point source pollution, Congress built NPS provisions into its 1987 amendments to the CWA. Since that time, federal, state, and local regulators, water quality managers, and government planners invested considerable time, money, and effort in developing programs to manage water quality problems caused by NPS pollution.

Nonpoint source pollution is a major - and extremely difficult - environmental problem to control. Recent studies by EPA and state water quality agencies indicate that the majority of the remaining water quality impairments to the nation's rivers, streams, lakes, estuaries, coastal waters, and wetlands result from NPS pollutants. EPA and the states together have developed NPS pollution management programs to tackle this difficult problem. In addition to NPS pollution controls under CWA Section 319, several other federal, state, and local programs supplement and interface with the CWA NPS pollution program.

NPS pollution is water pollution resulting from land runoff and seepage that occurs during rainfall and snowmelt. Precipitation causes runoff to pick up and carry naturally occurring pollutants and pollutants from human activities to surface and groundwater sources. Although diffuse runoff is generally treated as a NPS, runoff that enters and is discharged from certain conveyances, such as pipes, ditches, channels, tunnels, or other conduits, is considered a point source, which is subject to CWA permit requirements. For many reasons, nonpoint sources contribute significantly to the nation's water quality problems. Diffuse sources are generally harder to identify, isolate, and control than traditional point sources.

CWA Section 319(a), the principal federal regulatory vehicle for NPS pollution control, requires states to identify those waters impacted or threatened by nonpoint sources, and to identify the pollution sources affecting those water bodies. EPA's final report to Congress on Section 319 of the CWA summarizes and reviews the information provided by the states in their approved NPS assessments.[48] The uses most severely impacted by NPS pollution are wildlife (e.g., fish and waterfowl) and recreation (e.g., swimming). The nation's rivers, lakes, streams, estuaries, coastal waters, and wetlands are all experiencing major effects from NPS pollution. Agriculture is the single largest source of NPS pollution problems. Siltation and nutrients cause the greatest share of NPS impacts on U.S. surface waters.

Siltation is the major pollutant affecting rivers and streams. Nutrients, pathogens, metals, and pesticides are also present in varying degrees. Nationally, the greatest adverse impacts from siltation were found to occur in Missouri; from nutrients, North Dakota; from pathogens, Virginia; metals, Colorado; pesticides, Iowa; habitat modification, Montana; and organic sediments, North Carolina. The sources responsible for these adverse impacts to rivers and streams were (in order of impact)

[48] EPA-506/9-90 (Jan. 1992).

agriculture, surface mining, stream bank modification, on-site wastewater systems, subsurface mining, petroleum activities, channelization, flow regulation/modification, and urban runoff.

Nutrients (primarily nitrogen and phosphorous) are the greatest cause, and siltation the second major cause, of NPS pollution of lakes and ponds. Again, agricultural use constitutes the largest source of NPS pollution of lakes, followed by on-site wastewater systems, in-place contaminants, petroleum activities, flow regulation/modification, streambank modification, atmospheric deposition, urban growth, highway maintenance and runoff, recreation, and urban runoff.

Oil and grease constitute the primary NPS pollutant affecting coastal waters. Metals, pesticides, other inorganics, and pathogens are major NPS pollutants, while nutrients, siltation, and dissolved oxygen from nonpoint sources impact coastal waters to a lesser degree. Waste storage and storage tank releases account for the greatest share of NPS pollution of coastal waters, followed by impacts from mining and urban sources.

Nutrients are the greatest problem for estuaries, although pathogens cause the greatest amount of damage. Dissolved oxygen, oil and grease, salinity, priority organics, and metals rank high on the list of NPS pollution of estuaries. In-place contaminants are the biggest identified source of NPS pollution of estuaries, followed by on-site wastewater systems and petroleum activities. Urban runoff is also a large source of estuary pollution.

Understanding the various types of water bodies, types of pollutants, and sources of NPS pollution enables effective implementation of state NPS pollution control programs. When establishing planning measures to reduce the adverse impacts of NPS pollution, the information contained in the state's NPS assessments should be reviewed.

2.3.1 Clean Water Act Section 319

Recognizing the many nonpoint sources of pollution, and their interrelation to other aspects of water pollution control, Congress assigned EPA the role of establishing and coordinating a national NPS control program through Section 319 of the CWA. State Section 319 management programs provide a means to implement a consistent set of activities within each state to achieve state-established water quality objectives.

States are required to develop NPS management programs under CWA Section 319. Specifically, the management programs are required to:

* Identify the best management practices (BMPs) that will reduce pollutant loadings from nonpoint sources identified in state assessment reports;
* Identify programs to achieve implementation of BMPs including nonregulatory or regulatory programs for enforcement, technical assistance, education, training, technology transfer, and demonstration projects;
* Provide a schedule of annual milestones for the management program;
* Include certification by the state attorney general that the laws of the state provide adequate authority to implement the state's management program;
* Identify sources of federal and other financial support for NPS control activities; and
* Identify federal programs and projects that the state will review for consistency with the state NPS management program.

2.3.2 Coastal Nonpoint Source Pollution Controls

To combat nonpoint pollution of coastal waters, the Coastal Zone Act Reauthorization Amendments of 1990 require that each state with a federally approved Coastal Zone Management Program develop a "Coastal Nonpoint Pollution Control Protection Program" to implement coastal land use management measures for controlling NPS pollution. EPA has published national guidelines on "management measures" to control coastal nonpoint sources.[49]

Management measures are "economically achievable measures" for the control of pollutants from new and existing nonpoint sources that reflect the "greatest degree of pollutant reduction achievable" through application of the best available

[49]Guidance Specifying Management Measures for Sources of Nonpoint Pollution of Coastal Waters, 840-B-92-002 (Jan. 1993).

NPS control measures and methods. States are required to submit their proposed programs to the Secretary of Commerce and the EPA Administrator by July 1995. Following program approval, the states must implement their programs through changes to state management plans for the control of NPS pollution approved under CWA Section 319 and through changes to state Coastal Zone Management Programs. If a state fails to submit an approvable program, the Secretary of Commerce may withhold a percentage of any Coastal Zone Management Act Section 306 grant money, and EPA may withhold portions of any Section 319 grant under the CWA.

2.3.3 Other Federal NPS Pollution Control Programs

Several different agencies and laws govern control of NPS pollution. CWA Section 319 provides the primary framework for integrating all the various federal programs into state NPS management program. EPA and state water quality agencies have established water quality criteria and standards, directed construction of sewage treatment plants, national point source pollution control programs, wetland regulatory programs, and drinking water quality standards. EPA implements and oversees many different programs aimed at protecting and restoring water quality. An overview of some of these programs is provided here.

Clean Lakes Program. The Clean Lakes Program was created in 1976 to establish methods for protecting and restoring publicly owned freshwater lakes. Clean Lakes Program regulations were issued in 1980, and since then over 400 cooperative agreements have been awarded to states for the classification, assessment, study, and restoration of lakes. These agreements are subject to General Grant Regulations (40 C.F.R. Part 30) and Clean Lake Regulations (40 C.F.R. Part 35, Subpart H). Four types of cooperative agreements are available under the Clean Lakes Program. Under the Clean Lakes Regulations, all lake projects must be consistent with a state's Water Quality Management Plan to ensure that EPA and the states coordinate various programs under the CWA, Resource Conservation and Recovery Act, Safe Drinking Water Act, and other EPA programs.

Clean Lakes projects are developed and implemented on a watershed basis. Clean Lakes Program Guidance specifies that the watershed approach is the key element in successful management of NPS pollution, groundwater protection, water quality-based permitting, stormwater permitting, estuarine protection and cleanup, and wetlands protection. Many states have used the Clean Lakes Program as a NPS management tool. A number of projects have included BMPs to prevent pollutants originating in the watershed from entering lakes.

Rural Clean Water Program (RCWP). The RCWP provides financial and technical assistance to agricultural landowners and operators to test policies and procedures for controlling agricultural NPS pollution in 21 project areas. To aid in the development of BMPs, the Agricultural Stabilization and Conservation Service uses RCWP water quality data to evaluate BMPs.

Wellhead Protection Program. The 1986 Amendments to the Safe Drinking Water Act (SDWA) established a Wellhead Protection Program to protect groundwaters that contribute to public drinking water supplies. Under SDWA Section 1428, each state was required to prepare a Wellhead Protection Program for approval by EPA. To assist states in developing these programs, EPA has provided numerous technical assistance documents and has held national wellhead conferences and workshops. Those portions of the state wellhead protection programs linked to state NPS management programs can be used to support Section 319 activities to protect groundwater. If properly targeted, control efforts to protect wellfields can complement NPS efforts to prevent surface water degradation.

Municipal Pollution Control Programs. Municipal Pollution Control Programs manage efforts to assist communities in constructing new or upgraded municipal wastewater facilities needed to achieve compliance with federal surface water discharge standards, as well as to protect groundwater. Under Title VI of the 1987 Clean Water Act Amendments, the State Revolving Fund (SRF) program was developed to provide capitalization grants to states to establish SRFs. This program enables states to provide financial assistance to municipal wastewater facilities, as well as for "expanded uses," which includes NPS management. Municipal Pollution Control Programs has established the Expanded Uses Council, composed of directors from each of the EPA Office of Water program offices, to provide further guidance and coordination of SRF assistance for expanded uses.

National Estuary Program. EPA administers the National Estuary Program under Section 320 of the CWA to assist state, regional and local governments in developing comprehensive conservation and management plans for geographically targeted, high-priority estuarine waters. These plans recommend priority corrective actions to restore estuarine water quality, fish populations, and other designated uses of estuarine waters.

Pesticides Program. Under the Pesticides Program, EPA controls pesticides that may threaten ground and surface waters. EPA has developed a Pesticides and Groundwater Strategy pursuant to the Federal Insecticide, Fungicide and Rodenticide Act, which requires states to implement EPA-approved Pesticide State Management Plans to reduce and eliminate groundwater contamination from pesticide use.

Chapter 3

STORM WATER PERMIT PROCESS

3.1 OVERVIEW

This chapter describes the different options that facility operators may follow for obtaining permit coverage for storm water discharges associated with industrial activity:

1. Individual permit applications;
2. Group applications; and
3. General permit applications; or
4. Multisector permit coverage.

3.1.1 Types of Permits

The regulations present three permit application options for storm water discharges associated with industrial activity. The *first option* is to submit an individual application consisting of Forms 1 and 2F. The *second option* allows the applicant to become a participant in a group application. The *third option* is to file a Notice of Intent (NOI) to be covered under a general permit in accordance with the requirements of an issued general permit. In addition, on November 19, 1993, EPA issued proposed regulations governing new *multisector* storm water permits.[50] The proposed multisector permit would cover storm water discharges from 29 industrial sectors. The proposed multisector general permit was designed in response to the second of the three options listed above. EPA classified groups into 29 industrial sectors where the nature of industrial activity, type of materials handled, and material management practices employed were sufficiently similar for the purposes of developing multisector permit conditions. Each of these types of permit applications is discussed in this chapter.

3.1.2 Permit Application Process

This section provides a summary of the permit process for storm water discharges associated with industrial activity.

[50] 58 Fed. Reg. 61146 (Nov. 19, 1993).

Following the submission of the National Pollutant Discharge Elinination System (NPDES) permit application, the permitting authority reviews the application for completeness. If additional information is required to complete the application, the permitting authority will notify the applicant. The permitting authority will specify a deadline for submitting the additional information. The effective date of the application is the date when the permitting authority determines that the application is complete.

The permitting authority may request additional information beyond what is required in the application form. The permit writer will use available information, primarily that in the permit application, to develop a draft permit or a notice to deny a permit. All draft permits and notices of intent to deny a permit will include a statement of basis or a draft fact sheet. The statement of basis will briefly describe the rationale for either proceeding with issuing a permit or denying a permit. The draft fact sheet will include the principal facts, methodology, and any legal or policy questions considered in the decision to proceed with issuance of a permit.

All draft permits and notices of intent to deny a permit are subject to public notice and will be made available for public comment. The permitting agency will give public notice when:

1. A permit application has been tentatively denied,
2. A draft permit is issued,
3. An evidentiary hearing is granted, or
4. When a new source determination has been made.

After the close of the public comment period, the permitting agency will issue a final decision. The permitting agency, upon issuance of the final decision, will respond to comments, identify any changes in the tentative decision (to either permit or deny a permit) and give any reason pertinent to the changes. If a final NPDES permit is issued, the permit usually specifics the effective date, at which time the facility is legally authorized to discharge storm water associated with an industrial activity subject to the permit conditions.[51]

A permit will be issued by the permitting agency for a period up to, but not more than 5 years. Dischargers must reapply for a permit 180 days before the expiration date of the permit.

The permit is not transferable except after notice to and approval by the permitting authority. The Director of the permitting authority may require modification or revocation and reissuance of the permit to change the name of the permittee and incorporate such other requirements that may be necessary under the CWA.

CWA Section 402(j) requires that all permit applications be made available to the public. Information in permit applications is made available to the public upon request. Any information required in a permit application may not be claimed as confidential. Any information submitted to EPA which goes beyond that required by Form 1, Form 2F, or other appropriate forms may be claimed as confidential. However, claims for confidentiality of effluent data will be denied.

If a claim of confidentiality is not asserted at the time of submitting the information, EPA may make the information public without further notice to the applicant. Claims of confidentiality will be handled in accordance with EPA's business confidentiality regulations.

The CWA provides that any person who violates a permit condition is subject to a civil penalty not to exceed $25,000 per day of violation. Any person who willfully or negligently violates a permit is subject to a fine of not less than $2,500 or more than $25,000 per day of violation, or imprisonment for not more that 1 year, or both.[52]

The operator of a facility must allow a representative of the permitting authority upon the presentation of credentials and other documents as may be required by law, to enter the regulated facility and inspect records pertaining to the permit. This includes, but is not limited to, monitoring and control

[51] A more complete description of the processes involved in obtaining an NPDES permit is provided in 40 C.F.R. Part 124, especially Subpart D.
[52] 40 C.F.R. 122.41(a).

equipment, practices, and operations regulated under this permit. The representative may also sample the storm water discharge for any substance to ensure compliance with the permit conditions. Inspection activities are to be conducted at reasonable times.[53]

The operator must retain all records of discharge monitoring for at least three years from the date of the sample, measurement, report, or application. This includes all calibration and maintenance records, all original strip charts from continuous monitoring, copies of all records required by the permit, and all records of data used to complete the NPDES permit application.[54]

The CWA provides that any person who knowingly falsifies any record or document, tampers with or renders inaccurate any monitoring device, will be punished by a fine of not more than $10,000 per violation, or by imprisonment for not more than 2 years, or both.[55]

3.1.3 Permit Application Checklist

Prospective applicants seeking an NPDES permit for storm water-related industrial activity can refer to the following list that summarizes the applicant's primary responsibilities. This application checklist is useful for the applicant for self-checking the completeness of the application prior to submission. Applications will not be considered complete unless all applicable information required is provided. If additional information is required, the applicant will be notified.

1. Determine whether a permit is required for the storm water discharge.

   * Refer to this chapter
   * Contact the permitting authority, if necessary
   * Record name of contact person

[53] 40 C.F.R. 122.41(i)(1) to (4).

[54] 40 C.F.R. 122.41(j)(2).

[55] 40 C.F.R. 122.41(j)(5) and (k)(2).

2. Determine whether the state in which the discharge(s) is located has an EPA-approved NPDES program.

* Refer to Appendix B of this book
* Determine which forms need to be submitted for individual applications.
* If EPA is the permitting authority, list appropriate forms.
* For EPA-approved states, contact the permitting authority for appropriate forms and instructions.

3. Determine if a general permit will be, or has been, issued for the discharge.

4. If no general permit, select between participating in a group application or submitting an individual application.

5. Determine what the deadlines are for the permit application.

* Check Section 2.2.7 if EPA is the permitting authority
* Contact the state permitting agency if this information is not provided in the application form or instructions provided by that agency.

6. Complete the appropriate application forms. All applicants are to submit Forms #1 and 2F. Determine whether Forms 2C, 2D, and/or 2E need to be submitted.

7. Retain a complete copy of the permit application and all supporting documentation.

8. Submit the completed application forms to the appropriate permitting agency by the application deadline identified.

3.2 INDIVIDUAL PERMITS

Operators of facilities with storm water discharges associated with industrial activity who did not participate in a group application or did not obtain coverage under a general permit must submit an individual application consisting of Form 1 and Form 2F. The information required in Form 2F includes a site drainage map, a narrative description of the site identifying potential pollutant sources, and quantitative testing data. There are specific requirements for construction activities and oil and gas operations and mining operations.

This Section focuses on the procedures and requirements associated with submitting individual permit applications. Section 3.2.1 discusses the process of submitting individual permit applications. Section 3.2.2 provides an overview of the requirements of Form 1 and Form 2F, the individual permit application forms for most storm water discharges associated with industrial activity. Section 3.2.3 discusses special provisions for individual applications for storm water discharges associated with industrial activity from: small businesses; construction activities; and mining and oil and gas operations. Section 3.2.4 discusses deadlines for submitting individual permit applications. Section 3.2.5 describes the additional application forms that are necessary if storm water associated with industrial activity is mixed with nonstorm water. Section 3.2.6 explains where to obtain and submit permit applications. Section 3.2.7 describes signatory requirements for individual permit applications, and Section 3.2.8 describes penalties for knowingly submitting false information.

3.2.1 Process of Submitting Individual Applications

The process of selecting and submitting the application forms to use for individual permit applications for storm water discharges associated with industrial activity are as follows:

1. Determine whether the discharge is considered a storm water discharge associated with industrial activity. Refer to the definition of "storm water discharge associated with industrial activity" provided in 40 C.F.R. 122.2(33).

2. Determine whether the State in which the discharge is located has an authorized NPDES program. A list of these States is provided in Appendix B. Please note that the permit application forms required by authorized NPDES States may be different from the EPA-required forms that are discussed in this chapter.

3. Determine the track (e.g., individual permit application track, group application track, general permit track, etc.) that the discharger will pursue to comply with application requirements.

4. Obtain the appropriate application forms if submitting an individual permit application. Sections 3.2.2, 3.2.3, and 3.2.5 of this chapter provide information on permit application forms and requirements. Section 3.2.6 describes where forms can be obtained.

5. Submit the completed application to the appropriate permitting regulatory agency by the application deadline (Section 2.2.7). Section 3.2.6 describes where applications are to be submitted.

3.2.2 Forms 1 and 2F

The requirements for individual permit application for most types of discharges composed of storm water associated with industrial activity are incorporated into Form 1 and Form 2F.[56]

Form 1 (EPA Form 3510-1) requires general information about the facility, including: the name and address of the facility; the facility type (i.e., Standard Industrial Classification [SIC] code); a map showing specified features, etc.

[56]See Section 3.2.3 for a discussion of alternative individual permit application requirements for storm water discharges. In addition, Section 3.2.5 discusses the additional forms necessary where storm water discharges associated with industrial activity are mixed with any nonstorm water discharge.

Form 2F (EPA Form 3510-2F) contains information which can be used to evaluate the pollution potential of storm water discharges associated with industrial activity, including:[57]

* A map showing site drainage;
* An estimate of the area of impervious surfaces and the total area drained by each outfall;
* A narrative description of material management practices and control measures;
* A certification that separate storm water outfalls have been tested or evaluated for nonstorm water discharges;
* Existing information regarding significant leaks or spills of toxic or hazardous pollutants at the facility that have taken place within the three years prior to the submittal of the application; and
* Sampling data for specified parameters.

3.2.3 Special Provisions for Selected Discharges

Small businesses with storm water discharges associated with industrial activity do not have to analyze storm water discharges associated with industrial activity for the organic toxic pollutants listed in Table 2F-3 of Form 2F.[58]

There are two ways in which a facility can qualify as a "small business." If the facility is a coal mine, and if the probable total annual production is less than 100,000 tons per year, past production data or estimated future production (such as a schedule of estimated total production under 30 C.F.R. 79514(c)) may be submitted instead of conducting analyses for the organic toxic pollutants. Facilities that are not a coal mine with gross total annual sales for the most recent three years

---

[57] Chapter 4 provides technical guidance for obtaining or estimating the following information required by Form 2F: preparing a site drainage map, detecting the presence of nonstorm water discharges, measuring storm water runoff flow rates and volumes, and sampling equipment and procedures for collecting storm water discharge samples.

[58] Small businesses with storm water discharges associated with industrial activity are subject to the other appropriate requirements of Form 1 and Form 2F, including requirements to sample for specified conventional pollutants and other specified constituents. 40 C.F.R. 122.21(g)(8).

average less than $100,000 per year (in second quarter 1980 dollars) may submit sales data for those years instead of conducting analyses for the organic toxic pollutants. The production or sales data must be for the facility which is the source of the discharge. The data should not be limited to production or sales for the process or processes which contribute to the discharge, unless those are the only processes at the facility. For sales data, in situations involving intracorporate transfer of goods and services, the transfer price per unit should approximate market prices for those goods and services as closely as possible. Sales figures for years after 1980 should be indexed to the second quarter of 1980 by using the gross national product price deflator (second quarter of 1980 = 100).[59]

The application requirements for operators of storm water discharges associated with industrial activity from construction activities include Form 1 and a narrative description of:

1. The location (including a map) and the nature of the construction activity;

2. The total area of the site and the area of the site that is expected to undergo excavation during the life of the permit;

3. Proposed measures, including best management practices, to control pollutants in storm water discharges during construction, including a brief description of applicable State and local erosion and sediment control requirements;

4. Proposed measures to control pollutants in storm water discharges that will occur after construction operations have been completed, including a brief description of applicable State and local storm water management controls;

5. An estimate of the runoff coefficient of the site and the increase in impervious area after the construction addressed in the permit application is completed, the nature of fill

[59] This index is available in National Income and Product Accounts of the United States (Department of Commerce, Bureau of Economic Analysis).

material and existing data describing the soil or the quality of the discharge; and

6. The name of the receiving water.

At this time, EPA has not developed a standardized form for the narrative information accompanying Form 1 that is required in individual applications for storm water discharges associated with industrial activity from construction sites.

Several specific regulatory provisions are applicable to storm water discharges associated with industrial activity from mining and oil and gas operations:

1. *Mining Operations and Oil and Gas* (40 C.F.R. 122.26(a)(2)): The permitting authority may not require a permit for discharges of storm water runoff from mining operations or oil and gas exploration, production, processing, or treatment operations or transmission facilities composed entirely of flows which are from conveyances or systems of conveyances (including but not limited to pipes, conduits, ditches, and channels) used for collecting and conveying precipitation runoff and which are not contaminated by contact with or that has not come into contact with any overburden, raw material, intermediate products, finished product, by-product, or waste products located on the site of such operations.

2. *Oil and Gas* (40 C.F.R. 122.26(c)(1)(iii)): The operator of an existing or new discharge composed entirely of storm water from an oil or gas exploration, production, processing, or treatment operation or transmission facility is not required to submit a permit application, unless the facility:
   (a) has had a discharge of storm water resulting in the discharge of a reportable quantity for which notification is or was required pursuant to 40 C.F.R. 117.21 or 40 C.F.R. 302.6 at any time since November 16, 1987; or
   (b) has had a discharge of storm water resulting in the discharge of a reportable quantity for which notification is or was required pursuant to 40 C.F.R. 110.6 at any time since November 16, 1987; or

(c) contributes to a violation of a water quality standard.

3.2.4 Individual Applications Deadlines

Individual permit applications for storm water discharges associated with industrial activity which are currently not covered by an NPDES permit were to have been submitted by October 1, 1992.[60] Operators of discharges which are authorized by an individual NPDES permit must resubmit individual permit applications 180 days prior to the termination of the existing NPDES permit.

Permit applications for a new discharge of storm water associated with industrial activity must be submitted 180 days before that facility commences industrial activity which may result in a discharge of storm water associated with that industrial activity. Permit applications for a new discharge of storm water associated with industrial activity from a construction activity must be submitted at least 90 days before the date on which construction is to commence. Persons proposing a new discharge are encouraged to submit their application well in advance of the 90- or 180-day requirements to avoid delay.

Where a general permit has been issued, deadlines for submitting a NOI to be authorized to discharge under the permit are established in the permit.

3.2.5 Additional Forms

Where a storm water discharge associated with industrial activity is mixed with a nonstorm water component prior to discharge, an additional application form must be submitted.

A complete permit application for a storm water discharge associated with industrial activity mixed with process wastewater (process wastewater is water that comes into direct contact with or results from the production or use of any raw material, intermediate product, finished product, by-product, waste product or wastewater) includes Form 1, Form 2F, and Form 2C.

---

[60] See Section 2.2.7 and Table 2.2 for specific permit application deadlines.

A complete permit application for a storm water discharge associated with industrial activity mixed with new sources or new discharges of nonstorm water (non-NPDES permitted discharges commencing after August 13, 1979) includes Form 1, Form 2F, and Form 2D.

A complete permit application for a storm water discharge associated with industrial activity mixed with nonprocess wastewater (nonprocess wastewater includes noncontact cooling water and sanitary wastes which are not regulated by effluent guidelines or a new source performance standard, except discharges by educational, medical, or commercial chemical laboratories) includes Form 1, Form 2F, and Form 2E.

3.2.6 Where to Obtain and Submit Applications

In states without an authorized NPDES State program, EPA issues all NPDES permits. Where EPA issues permits, permit application forms can be obtained from and submitted to the appropriate EPA Regional office.[61] In States with authorized NPDES programs, application forms can be obtained from and submitted to the appropriate State office.[62] The permit application forms required by authorized NPDES States may be different from the EPA-required forms that are discussed in this chapter.

3.2.7 Signatories

Section X of Form 2F requires that all permit applications must be signed with the following certification:

> *"I certify under penalty of law that this document and all attachments were prepared under my direction or supervision in accordance with a system designed to assure that qualified personnel properly gather and evaluate the information submitted. Based on my inquiry of the person or persons who manage the system, or those persons directly responsible for gathering the information, the information submitted is, to the best of my knowledge and belief true, accurate, and complete.*

[61] Appendix B contains a list of the addresses and telephone numbers of the EPA Regional offices.

[62] A list of State offices is provided in Appendix B.

I am aware that there are significant penalties for submitting false information, including the possibility of fine and imprisonment for knowing violations."
This certification is to be signed as follows:

1. *For a corporation:* by a responsible corporate official. For purposes of this section, a responsible corporate official means:

(a) A president, secretary, treasurer, or vice-president of the corporation in charge of a principal business function, or any other person who performs similar policy or decision-making functions for the corporation, or

(b) The manager of one or more manufacturing, production, or operating facilities employing more than 250 persons or having gross annual sales or expenditures exceeding $25,000,000 (in second-quarter 1980 dollars), if authority to sign documents has been assigned or delegated to the manager in accordance with corporate procedures.

EPA does not require specific assignments or delegation of authority to responsible corporate officers. The Agency will presume that these responsible corporate officers have the requisite authority to sign permit applications unless the corporation has notified the Director to the contrary. Corporate procedures governing authority to sign permit applications may provide for assignment or delegation to applicable corporate position rather than to specific individuals.

2. *For a partnership or sole proprietorship:* by a general partner or the proprietor, respectively; or

3. *For a municipality, State, Federal, or other public agency:* by either a principal executive officer or ranking elected official. For purposes of this section, a principal executive officer of a Federal agency includes:

(a) The chief executive officer of the agency, or

(b) A senior executive officer having responsibility for the overall operations of a principal geographic unit of the agency (e.g. Regional Administrators of EPA).

3.2.8 Penalties For Knowingly Submitting False Information

The CWA provides for severe penalties for knowingly submitting false information on application forms. CWA Section 309(c)(4) provides that:

"Any person who knowingly makes any false material statement, representation, or certification in any application, . . . shall upon conviction, be punished by a fine of not more than $10,000 or by imprisonment for not more than 2 years or by both. If a conviction of such person is for a violation committed after a first conviction of such person under this paragraph, punishment shall be by a fine of not more than $20,000 per day of violation, or by imprisonment of not more than 4 years or by both."

3.2.9 Summary of Individual Application Requirements

These requirements address storm water discharges associated with industrial activity that are not authorized by a general permit and that are not included in a group application.

Application Forms

[] Applicants for discharges composed entirely of storm water must submit Forms 1 and 2F

[] Applicants for discharges composed of storm water and process wastewater must submit Forms 1, 2C, and 2F

[] Applicants for new sources or new discharges composed of storm water and nonstorm water must submit Forms 1, 2D, and 2F

[] Applicants for discharges composed of storm water and nonprocess wastewater must submit Forms 1, 2E, and 2F

[] Authorized NPDES States may establish their own forms which are at least as stringent as EPA's forms.

[] Forms are available from State permitting authorities for facilities located in NPDES authorized States, or from EPA Regional Offices for facilities located in States without NPDES authorization.

Form 2F Requirements

[] Site map showing topography and/or drainage areas and site characteristics.

[] Estimate of impervious surface area and the total area drained by each outfall.

[] Description of significant materials exposed to storm water, including current materials management practices.

[] Certification that outfalls have been tested or evaluated for the presence of nonstorm water discharges that are not covered by a NPDES permit.

[] Information on significant leaks and spills in the last 3 years.

[] Quantitative testing data for the following parameters:

- Any pollutants limited in an effluent guideline to which the facility is subject
- Any pollutant listed in the facility's NPDES permit for process wastewater
- Oil and grease, pH, biochemical oxygen demand (BOD_5), chemical oxygen demand (COD), total suspended solids (TSS), total phosphorus, nitrate plus nitrite nitrogen, and total Kjeldahl nitrogen
- Certain pollutants known to be in the discharge
- Flow measurements or estimates
- Date and duration of storm event.

Individual Application Requirements for Construction Activities

[] Provide a narrative description of:

- Location and nature of construction activity (including a map)
- Total area of the site and area to be excavated
- Proposed measures to control pollutants in storm water discharges during and after construction operations
- Estimate of runoff coefficient and increase in impervious areas after construction

- Name of receiving water.
[] No quantitative sampling.
[] Application deadline
- 90 days prior to date when construction begins.
[] EPA has not developed a standard form for these discharges at this time (Form 2F is not required).

Application Requirements for Oil and Gas Operations and Mining Operations

[] Operators of oil and gas facilities are not required to submit a permit application unless the facility:
- Has had a discharge of a reportable quantity for which notice is required under the Comprehensive Environmental Response, Compensation, and Liability Act (CERCLA) or the CWA in the past 3 years, or
- Contributes to a violation of a water quality standard.
[] Operators of active and inactive mining sites are not required to submit permit applications unless the discharge has come into contact with any overburden, raw material, intermediate or finished products, by-products, or waste products located onsite (inactive coal mining operations released from Surface Mining Coal Reclamation Act (SMCRA) performance bonds and noncoal mining operations released from applicable State or Federal reclamation requirements after December 17, 1990, are not required to submit permit applications).

Deadline

[] October 1, 1992, or 180 days prior to commencement of a new discharge.

3.3 GROUP PERMITS

The group application procedure is an option available for facilities that are part of the same effluent guideline subcategory or have similar industrial operations, waste streams, and other characteristics. Group applications reduce the burden on the regulated community by requiring the submission of quantitative data from only selected members of the group.

The group application is submitted in two parts. Part 1 of the application identifies all participants, provides facility-specific information, and proposes a representative sampling subgroup. Part 2 of the application consists of sampling data from each member of the sampling subgroup identified in Part 1 of the application.[63]

Part 1 of the group application was due by September 30, 1991, and Part 2 of the application was due by October 1, 1992. These deadlines applied to all industrial activities except those owned or operated by a municipality with a population of less than 250,000. For these facilities, Part 1 of the application was due by May 18, 1992, and Part 2 of the application was due by May 17, 1993. Both Parts were submitted directly to U.S. EPA Headquarters, Office of Wastewater Enforcement and Compliance (EN-336), 401 M Street, SW, Washington, DC 20460, regardless of whether the included facilities were in a NPDES-authorized State.

The Intermodal Surface Transportation Act also addressed municipally owned or operated industrial activities that were denied EPA approval for the group application process. Such facilities had to submit an individual application or be covered by a general permit within 180 days after the denial was made, or by October 1, 1992, whichever was later.

EPA formulates model permit language for group applications. Completed applications and model permit language are then distributed to every NPDES-authorized State or EPA Region (if the State is not NPDES authorized) in which participants are located. The State then reviews the application and model permit language. The State may consider the application and model permit language when issuing permits (either individual or general). The State may ask each or any of the applicants for more information on their facility and/or discharge if the State needs additional information. EPA Regional Offices will follow these same steps for participants located in States without NPDES authorization.

---

[63] A model group application accompanied by detailed information on how to complete both Part 1 and Part 2 group applications is available from the Storm Water Hotline, (703) 821-4823. Technical support with regard to sampling procedures is also available from the hotline.

3.3.1 Part 1 Requirements

The following information is required under Part 1 of the Group Applications:

[] A list of participants by name, location, and precipitation zone
[] A summary of each participant's industrial activities
[] An explanation of why the participants are sufficiently similar
[] A list of significant materials stored outside by each participant and materials management practices.
[] A list of representative dischargers that will submit test data in Part 2.

3.3.2 Part 2 Requirements

Under Part 2 of the Group Applications, quantitative testing data must be submitted by those facilities identified as "samplers" in Part 1 of the application. For groups of 4 to 20 members, 50% of the facilities must submit data; for groups with 21 to 99, a minimum of 10 dischargers must submit quantitative data; for groups with 100 to 1,000 members, a minimum of 10% of the facilities must submit data; for groups with greater than 1,000 members, no more than 100 facilities must submit data; there must be two dischargers from each precipitation zone in which ten or more members of the group are located, or one discharger from each precipitation zone in which nine or fewer members are located. Testing requirements are described under 40 C.F.R. 122.26(c)(1)(i)(E) and 40 C.F.R. 122.21 (g)(7).

3.3.3 Multisector Permits

On November 19, 1993, EPA issued proposed regulations governing new multisector storm water permits.[64] The proposed multisector permit would cover storm water discharges associated with industrial activity to U.S. waters, including discharges through large and medium municipal separate storm sewer systems, and through other municipal separate storm

[64] 58 Fed. Reg. 61146 (Nov. 19, 1993).

sewer systems. The proposed permit would cover discharges from 29 industrial sectors. The EPA Notice in the Federal Register provides draft general permits and fact sheets with industry-specific sections that describe Storm Water Pollution Prevention Plan requirements, numeric effluent limitation requirements, and monitoring requirements for each industry.

The regulations presented three permit application options for storm water discharges associated with industrial activity. The first option was to submit an individual application consisting of Forms 1 and 2F.[65] The second option was to become a participant in a group application.[66] The third option was to file a NOI to be covered under a general permit in accordance with the requirements of an issued general permit.[67]

The proposed multisector general permit was designed in response to the second of these three options. In Part 1 of the group applications, all participants were identified and information on each facility was included, such as industrial activities, significant materials exposed to storm water, and material management activities. For Part 1 of the application, groups also identified sampling subgroups to submit sampling data for Part 2 of the group application.

Over 1,200 groups with over 60,000 member facilities submitted Part 1 applications. Upon review of the Part 1 application, if the EPA determined that the application was an appropriate grouping of facilities with complete information provided on each participant and a suitable sampling subgroup was proposed, the application was approved. Part 2 of the application consisted of sampling data from each member of the sampling subgroup identified in Part 1 of the application.

In drafting the proposed multisector general permit, EPA reviewed both parts of the applications and formulated proposed permit language. To facilitate the process of developing permit conditions for each of the 1200 group applications submitted, EPA classified groups into 29 industrial sectors where the nature of industrial activity, type of materials handled, and material management practices employed were sufficiently similar for the

[65] See Section 3.2.
[66] See Section 3.3.
[67] See Section 3.4.

purposes of developing permit conditions. Each of the industrial sectors were represented by one or more groups which participated in the group application process.

3.4 GENERAL PERMITS

On September 9th and 25th, 1992, EPA issued general permits for construction and industrial activities[68] which are intended to initially cover the majority of storm water discharges associated with industrial activity in 12 states and 6 territories without authorized NPDES programs. Facilities in authorized NPDES States should contact their State permitting agencies to determine the status of the general permitting program. This section outlines conditions in EPA's general permits for industrial activities and construction activities.

3.4.1 Areas of Coverage

The following is a breakdown by EPA Region of the areas covered by the general permit requirements:

Region I: MA, ME, NH; Indian lands in MA, NH, ME.
Region II: PR and Indian lands in NY.
Region III: DC, Federal facilities in DE.
Region IV: FL; Indian lands in FL, MS, NC.
Region VI: LA, NM, OK, TX.
Region VII: SD; Indian lands in CO, MT, ND, SD, UT (except Goshute Reservation and Navajo Reservation lands), WY; Federal facilities in CO; Ute Mountain Reservation in CO, and NM.
Region IX: American Samoa and Guam; AZ; Territories of Johnston Atoll, and Midway and Wake Island; Indian lands in CA, and NV; Goshute Reservations in UT and NV, Navajo Reservations in UT, NM, and AZ, Duck Valley Reservation in NV and ID.
Region X: AK, and ID; Indian lands in AK, ID (except Duck Valley Reservation lands), and WA; Federal facilities in WA.

68See 57 Fed. Reg. 41176 (Sept. 9, 1992); 57 Fed. Reg. 44412 (Sept. 25, 1992).

3.4.2 Types of Discharges Covered

EPA's general permits cover the majority of storm water discharges associated with industrial activity. Storm water discharges associated with industrial activity that cannot be authorized by EPA's general permits include those:

* With an existing effluent limitations guideline for storm water

* That are mixed with nonstorm water, unless the nonstorm water discharges are in compliance with a different NPDES permit

* With an existing NPDES individual or general permit for the storm water discharges

* That are or may reasonably be expected to be contributing to a violation of a water quality standard

* That are likely to adversely effect a listed or proposed to be listed endangered or threatened species or its critical habitat

* From inactive mining, or inactive oil and gas operations, or inactive landfills occurring on Federal lands where an operator cannot be identified (industrial permit only).

3.4.3 Special Conditions

The following special conditions are applicable to general permits:

* Prohibition on most types of nonstorm water discharges as a component of discharges authorized by this permit. (These discharges should already have an NPDES permit.) However, EPA's permits authorize certain types of nonstorm water discharges.

* In the event there is a release(s) of a hazardous substance in excess of reportable quantities, as established under the

CWA or CERCLA (see 40 C.F.R. 117.3, 40 C.F.R. 302.4), the discharger must notify the National Response Center and the Director and modify the Storm Water Pollution Prevention Plan.

3.4.4 NOI Requirements

Industrial storm water dischargers that submit an NOI to be covered by the general permit are not required to submit an individual permit application or participate in a group application, provided the discharger is eligible for the permit and an individual permit application is not required by the Director on a case-by-case basis. Submitting an NOI represents significantly less burden than submitting an individual application or participating in a group application. The NOI requirements for general permits usually address only general information and typically do not require the collection of monitoring data. Submittal of an NOI is only possible where applicable general permits have been issued by the permitting authority. The following is a list of the NOI requirements for general permits:

* A facility must submit an NOI to be authorized by the general permit.

* NOIs do not require the collection of discharge sampling data.

* Facilities which discharge to a large or medium municipal separate storm sewer system must also submit signed copies of the NOI to the operator of the municipal system.

* Operators of construction activities must also submit signed copies of the NOI to State or local agencies approving sediment and erosion or storm water management plans under which the construction activity is operating.

The deadlines for NOIs were as follows:

* On or before October 1, 1992 for existing industrial activities

* For facilities or construction activities which begin industrial activity after October 1, 1992, an NOI shall be submitted at least 2 days prior to the commencement of the industrial activity.

* NOIs must be sent to the following address:

Storm Water Notice of Intent
P.O. Box 1215
Newington, VA 22122

3.4.5 Contents of NOI for Industrial Activities

The specific information of the NOI requirements for industrial facilities that are submitting a general permit are as follows:

1. Street address or latitude/longitude.

2. SIC code or identification of industrial activity.

3. Operator's name, address, telephone number, and status as Federal, State, private, public, or other entity.

4. Permit number(s) of any existing NPDES permit(s).

5. Name of receiving water(s).

6. Indication of whether the owner or operator has existing quantitative data describing the concentration of pollutants in storm water discharges.

7. A certification that a Storm Water Pollution Prevention Plan has been prepared for the facility (for industrial activities that begin operations after October 1, 1992).

3.4.6 Pollution Prevention Plan Requirements for Industrial Activities

Operators of all facilities covered by EPA's general permits must prepare and implement a Storm Water Pollution Prevention Plan. The Pollution Prevention Plan is considered to be the most important requirement of the general permit. Each industrial facility covered by the general permit must develop a plan, tailored to the site-specific conditions and designed with the goal to control the amount of pollutants in storm water discharges from the site. The following checklist summarizes the pollution prevention plan requirements for industrial facilities covered by a general permit:

[] **Pollution Prevention Team** - Each facility will select a Pollution Prevention Team from its staff, and the team will be responsible for developing and implementing the plan.

[] **Components of the Plan** - The permit requires that the plan contain a description of potential pollutant sources and a description of the measures and controls to prevent or minimize pollution of storm water. The description of potential pollutant sources must include:

- A map of the facility indicating the areas which drain to each storm water discharge point.

- An indication of the industrial activities which occur in each drainage area.

- A prediction of the pollutants which are likely to be present in the storm water.

- A description of the likely source of pollutants from the site.

- An inventory of the materials which may be exposed to storm water.

- The history of spills or leaks of toxic or hazardous materials for the past 3 years.

The measures and controls to prevent or minimize pollution of storm water must include:

- Good housekeeping or upkeep of industrial areas exposed to storm water.

- Preventive maintenance of storm water controls and other facility equipment.

- Spill prevention and response procedures to minimize the potential for and the impact of spills.

- Test all outfalls to ensure there are no cross connections (only storm water is discharged).

- Training of employees on pollution prevention measures and controls, and recordkeeping.

The permit also requires that facilities:

- Identify areas with a high potential for erosion and the stabilization measures or structural controls to be used to limit erosion in these areas.

- Implement traditional storm water management measures (oil/water separators, vegetative swales, detention ponds, etc.) where they are appropriate for the site.

[] **Inspection/Site Compliance Evaluation:**
Facility personnel must inspect the plant equipment and industrial areas on a regular basis. At least once every year a more thorough site compliance evaluation must be performed by facility personnel to:

- Look for evidence of pollutants entering the drainage system.

- Evaluate the performance of pollution prevention measures.

- Identify areas where the plan should be revised to reduce the discharge of pollutants.

- Document both the routine inspections and the annual site compliance evaluation in a report.

[] **Consistency** - The plan can incorporate other plans which a facility may have already prepared for other permits including Spill Prevention Control and Countermeasure (SPCC) Plans, or Best Management Practice (BMP) Programs.

[] **Deadlines** - The plan must be prepared on or before April 1, 1993, and the facility must be in compliance with the plan on or before October 1, 1993.

[] **Signature** - The plan must be signed by a responsible corporate official, such as the president, vice-president, or general partner.

[] **Plan Review** - The plan is to be kept at the permitted facility at all times. The plan should be submitted for review only when requested by EPA.

[] **Semiannual Monitoring/Annual Reporting** is required of the following facilities:

- EPCRA Section 313 facilities.

- Primary metal industries, SIC 33.

- Land disposal units / incinerators / boilers and industrial furnaces (BIFs).

- Wood treatment facilities.

- Facilities with coal pile runoff.

- Battery reclaimers.

[] **Annual Monitoring** is required of the following types of facilities (although there are no reporting requirements for these facilities):

- Airports with at least 50,000 flight operations per year.

- Coal-fired steam electric facilities.

- Animal handling/meat-packing facilities.

- Additional facilities, including:

* SIC 30 and 28 with storage piles for solid chemicals used as raw materials that are exposed to precipitation.

* Certain automobile junkyards.

* Lime manufacturing facilities where storm water comes into contact with lime storage piles.

* Oil-handling sites at oil-fired steam electric power generating facilities.

* Cement manufacturing and cement kilns.

* Ready-mix concrete facilities.

* Shipbuilding and repairing facilities.

[] **Additional Monitoring Requirements**

- Testing parameters for facilities are listed in the general permits.

- At a minimum, all dischargers must conduct an annual site inspection of the facility.

[] **Alternative Certification**

- A discharger is not subject to the monitoring requirements for a given outfall if there is no exposure of industrial areas or activities to storm water within the drainage area of that outfall within a given year.

- The discharger must certify, on an annual basis, that there is no exposure to storm water, and such certification must be retained in the storm water pollution prevention plan. Facilities subject to semiannual monitoring requirements must submit this certification to EPA in lieu of monitoring data.

[] **Numeric Effluent Limitations**

- Coal pile runoff: 50 mg/l TSS and 6-9 pH.

3.4.7 Construction General Permit Requirements

Storm water discharges from construction sites that are authorized by general permits include those that will result in the disturbance of five or more acres of land.

3.4.8 Contents of NOI for Construction Activities

The specific information of the NOI requirements for construction sites that are submitting a general permit are as follows:

1. Street address or latitude/longitude.

2. The name, address, telephone number of the operator(s) with day to day operational control, and operator status as Federal, State, private, public, or other entity.

3. Permit number(s) of any existing NPDES permit(s).

4. Name of receiving water(s).

5. Indication of whether the owner or operator has existing quantitative data describing the concentration of pollutants in storm water discharges.

6. An estimate of the project start date and completion dates and estimates of the number of disturbed acres.

7. A certification that a Storm Water Pollution Prevention Plan has been prepared for the facility.

8. *Deadline for Notification:*
An NOI shall be submitted at least two days prior to the commencement of construction (commencement of construction is defined as the initial disturbance of soils associated with clearing, grading, or excavating activities or other construction activities) at any site that will result in the disturbance of five or more acres total land area.

3.4.9 Pollution Prevention Plan Requirements for Construction Activities

The Pollution Prevention Plan is considered to be the most important requirement of the general permit. Each construction activity covered by the general permit must develop a plan, tailored to the site-specific conditions and designed with the goal to control the amount of pollutants in storm water discharges from the site. The following checklist summarizes the pollution prevention plan requirements for construction activities covered by a general permit:

[] **Components of the Plan** - The permit requires that the plan contain a site description and a description of the measures and controls to prevent or minimize pollution of storm water. The site description must include:

- A description of the nature of the construction activity.

- A sequence of major construction activities.

- An estimate of the total area of the site and of the area to be disturbed.

- An estimate of the runoff coefficient of the site after construction is complete.

- Any existing data on the quality of storm water discharge from the site.

- The name of the receiving water.

- Any information on the type of soils at the site.

- A site map indicating drainage patterns and slopes after grading activities are complete, areas of soil disturbance, the outline of the area to be disturbed, the location of stabilization measures and controls, and surface waters at the discharge points.

[] **Measures and Controls** - Measures and controls to prevent or minimize pollution of storm water must include three different types of controls: erosion and sediment controls, storm water management controls, and other controls:

- *Erosion and Sediment Controls*:
* <u>Stabilization</u> (seeding, mulching, etc.) - disturbed areas where construction has permanently or temporarily ceased must be stabilized within 14 days of the last disturbance or as soon as practicable in semiarid and arid areas. (Areas which will be redisturbed within 21 days do not have to be stabilized.)

* <u>Structural Controls</u> - sites with common drainage locations that serve 10 or more disturbed acres must install a sediment basin where it is attainable (where a basin is not attainable, sediment traps, silt fence, or other equivalent measures must be installed. Sediment basins must provide 3,600 cubic feet of storage per acre drained. Drainage locations which serve less than 10

disturbed acres must install either a sediment basin, sediment trap, or silt fence along the down slope and side slope perimeter.

- The plan shall be completed prior to submittal of an NOI and updated as appropriate.

- For construction activities that have begun after October 1, 1992, the plan shall provide for compliance with the terms and schedule of the plan beginning with the initiation of construction activities.

3.4.10 Municipal Application Requirements

The CWA requires that NPDES permits for discharges from municipal separate storm sewer systems include a requirement to effectively prohibit nonstorm water discharges into the storm sewers, and controls to reduce the discharge of pollutants to the maximum extent practicable (including management practices, control techniques and system design and engineering methods, and other provisions appropriate for the control of such pollutants). EPA or authorized NPDES States may issue system-wide or jurisdiction-wide permits covering all discharges from a municipal separate storm sewer system. The November 1990 storm water final rule established requirements for a two-part permit application designed to facilitate development of site-specific permit conditions. The permit application requirements provide municipal applicants an opportunity to propose appropriate management programs to control pollutants in discharges from their municipal systems. This increases flexibility to develop appropriate permit conditions and ensures input from municipalities in developing appropriate controls. The general permit applications requirements for covered municipalities are as follows:[69]

[69] See also Guidance Manual for the Preparation of Part 1 of the NPDES Permit Application for Discharges from Municipal Separate Storm Sewer Systems, available from NTIS (703) 487-4650, order number PB 92-114578; Guidance Manual for the Preparation of Part 2 of the NPDES Permit

Part 1 Application Requirements

1. General information (name, address, etc.).

2. Existing legal authority and any additional authorities needed.

3. Source identification information.

4. Discharge characterization including:

   - Monthly mean rain- and snowfall estimates.

   - Existing quantitative data on volume and quality of storm water discharges.

   - A list of receiving water bodies and existing information on the impacts of receiving waters.

   - Field screening analysis for illicit connections and illegal dumping.

5. Characterization plan identifying representative outfalls for further sampling in Part 2.

6. Description of existing management programs to control pollutants from the municipal separate storm sewer and to identify illicit connections.

7. Description of financial budget and resources currently available to complete Part 2.

Applications for Discharges from Municipal Separate Storm Sewers Systems, available from the Storm Water Hotline, (703) 821-4823.

Part 2 Application Requirements

1. Demonstration of adequate legal authority to control discharges, prohibit illicit discharges, require compliance, and carry out inspections, surveillance, and monitoring.

2. Source identification indicating the location of any major outfalls and identifying facilities that discharge storm water associated with industrial activity through the municipal separate storm sewer.

3. Discharge characterization data including:

   - Quantitative data from five to ten representative locations in approved sampling plans.

   - For selected conventional pollutants and heavy metals, estimates of the annual pollutant load and event mean concentration of system discharges.

   - Proposed schedule to provide estimates of seasonal pollutant loads and the mean concentration for certain detected constituents in a representative storm event.

   - Proposed monitoring program for representative data collection.

4. Proposed management program including descriptions of:

   - Structural and source control measures that are to be implemented to reduce pollutants in runoff from commercial and residential areas.

   - Program to detect and remove illicit discharges.

   - Program to monitor and control pollutants from municipal landfills, hazardous waste treatment, disposal, and recovery facilities; EPCRA Section 313 facilities; and other priority industrial facilities.

- Program to control pollutants in construction site runoff.

5. Estimated reduction in loadings of pollutants as a result of the management program.

6. Fiscal analysis of necessary capital and operation and maintenance expenditures.

3.5 POLLUTION PREVENTION PLAN REQUIREMENTS

For each of the three types of storm water permits (i.e., individual, group, general), a Storm Water Pollution Prevention Plan must be developed for each facility covered by the permit. Storm Water Pollution Prevention Plans must be prepared in accordance with good engineering practices and in accordance with the factors outlined in 40 C.F.R. 125.3(d) (2) or (3), as appropriate. The plan must identify potential sources of pollution which may reasonably be expected to affect the quality of storm water discharges associated with activities at the facility. In addition, the plan must describe and ensure the implementation of practices which are to be used to reduce the pollutants in storm water discharges associated with activities at the facility and to ensure compliance with the terms and conditions of the permit. Facilities must implement the provisions of the Storm Water Pollution Prevention Plan as a condition of permit issuance. Specific pollution prevention plan requirements are discussed in Chapter 5.

Chapter 4

TECHNICAL GUIDANCE FOR STORM WATER PERMIT APPLICATIONS

4.1 STORM WATER SAMPLING

The storm water permit applications discussed in Chapter 3 have specific storm water sampling requirements. Sampling data gathered for the application will be used to characterize storm water discharges, and will serve as a basis for establishing requirements in National Pollutant Discharge Elimination System (NPDES) storm water permits. It is important to note that the applicant must report data that are representative of the storm water discharge and that the intentional misrepresentation of discharge characteristics is unlawful.

This chapter first presents a description of the fundamentals of storm water sampling and then provides additional technical guidance for obtaining the information required by Form 2F, including guidance for: developing site maps; identification of outfalls that discharge storm water associated with industrial activity; testing for the presence of nonstorm water discharges; estimating storm water runoff flow rates and volumes; and collecting samples.

4.1.1 Benefits of Sampling

Data that characterize storm water discharges are valuable to permitting authorities and permittees for several reasons. First, storm water sampling provides a means for evaluating the environmental risk of the storm water discharge by identifying the types and amounts of pollutants present. Evaluating these data helps to determine the relative potential for the storm water discharge to contribute to water quality impacts or water quality standard violations. Further, storm water sampling data can be used to identify potential sources of pollutants. These sources can then be either eliminated or controlled more specifically by the permit.

4.1.2 Industrial Facility Application Requirements

The storm water permit application regulations provide operators of facilities (including those owned by the government) that have storm water discharges associated with industrial activity with three application options:

1. Submit an individual application;
2. Participate in a group application (a two-part application); or
3. Submit a Notice of Intent (NOI) to be covered by a general permit where general permits are available.

This chapter focuses on sampling requirements for individual applications and Part 2 of group applications. Sampling data generally will not be required for an NOI; however, the general permit may require sampling during the term of the permit. State permitting authorities may also require sampling information for an NOI at their discretion and should, therefore, be consulted prior to submittal.

Industrial facilities submitting individual applications must submit sampling data on a completed application Form 2F (entitled "Application for Permit to Discharge Storm Water Discharges Associated with Industrial Activity"). Facilities selected to be part of the sampling subgroup for a group application must submit sampling data with Part 2 of the application. Members of the sampling subgroup must complete only the quantitative data portions of Form 2F, including

Sections VII, VIII, and IX, and the certification in Section X. Figure 4-1 details the types of information required for each section of Form 2F. Figure 4-2 describes what sampling information must be provided in Part 2 of the group application. It should be noted that States may require the use of different forms and submittal of additional documentation.

Form 1 must also be submitted with Form 2F by applicants submitting individual permit applications. General information about the facility is provided on Form 1 (i.e., addresses, operators, etc.); it does not request sampling data.

Facilities with unpermitted combined discharges of storm water and process or nonprocess wastewater must submit Form 2C or 2E, respectively, in addition to Forms 1 and 2F. Facilities with storm water discharges combined with new sources or new discharges of process wastewater must submit Form 2D, as well as Forms 1 and 2F.

Storm water discharge permit applications are generally submitted directly to the permit-issuing authority. The appropriate authority is the State, where the State has been granted the authority to issue NPDES permits, or the EPA Regional office where the state does not have NPDES authorization.

It should be noted, however, that both parts of a group application must instead be submitted to EPA Headquarters. Group applications must be sent to: Director, Office of Wastewater Enforcement and Compliance, U.S. EPA, EN-336, 401 M Street, SW, Washington, DC 20640.

Applications submitted by industrial facilities must be certified by a responsible corporate officer as described in 40 C.F.R. 122.22 (e.g., president, secretary, treasurer, vice-president of the corporation in charge of a principal business function). Applications submitted by municipalities must be certified by a principal executive officer or ranking elected official as described in 40 C.F.R. 122.22.

Figure 4-1
Form 2F Application Requirements

Section	Requirement
2F-I	Outfall location(s), including longitude and latitude and receiving water(s)
2F-II	Facility improvements which may affect the discharges described in the application
2F-III	Site drainage map I
2F-IVA	Estimates of impervious area within each outfall drainage area
2F-IVB	A narrative description of pollutant sources (i.e., onsite materials which may come in contact with storm water runoff)
2F-IVC	Location and description of existing structural and nonstructural pollutant control measures
2F-VA	Certification that outfalls have been tested or evaluated for nonstorm water discharges
2F-VB	Description of method used for testing/ evaluating presence of nonstorm water discharges
2F-VI	History of significant leaks or spills of toxic or hazardous pollutants at the facility within the last 3 years
2F-VII	Discharge characterization for all required pollutants
2F-VIII	Statement of whether biological testing for acute or chronic toxicity was performed and list of pollutants for which it was performed
2F-IX	Information on contract laboratories or consulting firms
2F-X	Certification that information supplied is accurate and complete

Figure 4-2
Part 2 Group Application Sampling Requirements

Quantitative Testing Data

[] For groups with 4 to 20 members, 50% of the facilities must submit data; for groups with 21 to 99 members, a minimum of 10 dischargers must submit quantitative data; for groups with 100 to 1,000 members, a minimum of 10% of the facilities must submit data; for groups with greater than 1,000 members, no more than 100 facilities must submit data; there must be 2 dischargers from each precipitation zone in which 10 or more members of the group are located, or 1 discharger from each precipitation zone in which 9 or fewer members are located.

[] Sampling and analysis requirements are described in 40 C.F.R. 122.26(c)(1)(i)(E) and 40 C.F.R. 122.21(g)(7). Pollutants to be analyzed depend on the type(s) of industries applying as a group.

[] Sampling subgroup must provide all quantitative discharge information required in Form 2F Sections VII-IX plus the certification in Section X.

[] The group application sampling subgroup must collect grab samples during the first 30 minutes of the storm event and flow-weighted composite samples as required in 40 C.F.R. 122.21(g)(7).

4.1.3 Municipal Application Requirements

Operators of large and medium municipal separate storm sewer systems are required to submit a twopart application. Both parts contain sampling requirements: Part 1 requires information characterizing discharges from the separate storm sewer system, including field screening sample data for identifying illicit/illegal connections; Part 2 requires sampling at representative locations and estimates of pollutant loadings for those sites. These sampling data are to be used to design a long-term storm water monitoring plan that will be implemented during the term of the permit. The sampling data that must be submitted in Parts 1 and 2 of municipal applications are listed in Figure 4-3. There is no standard application form for municipalities.

Figure 4-3
Municipal Applications Sampling Requirements

Part 1

[] Monthly mean rainfall and snowfall estimates

[] Existing quantitative data on the depth and quality of storm water discharges

[] A list of receiving water bodies and existing information concerning known water quality impacts

[] Field screening analysis for illicit connections and illegal dumping

[] Identification of representative outfalls for further sampling in Part 2

Part 2

[] Quantitative data from 5 to 10 representative locations in approved sampling plans

[] Estimates of the annual pollutant load and event mean concentration (EMC) of system discharges

[] Proposed schedule to provide estimates of seasonal pollutant loads and the EMC for certain detected constituents in a representative storm event during the term of the permit

[] Proposed monitoring program for representative data collection during the term of the permit

4.1.4 Who Must Sample

Operators of facilities that have storm water discharges associated with industrial activity and operators of large- and medium-size municipalities are required to conduct storm water sampling as part of their NPDES permit applications. Specifically, the following types of industries and municipalities must sample storm water discharges:

[] *Storm Water Discharges Associated with Industrial Activities*: See Table 2.3 in Chapter 2 for a listing of the 11 categories of facilities considered to be "engaging in industrial activity"; and

[] *Municipal Separate Storm Sewer Systems*: From a municipal separate storm sewer system serving a population of 100,000 or more, but less than 250,000.[70] Regulated municipalities are listed in Appendices F through I in 55 Fed. Reg. 47,990 (Nov. 16, 1990).

4.1.5 When Industrial Facilities Must Sample

Industrial individual and group applicants must include sampling data from at least one representative storm event. Operators of large or medium municipal separate storm sewer systems must submit sampling data from three different representative storm events. How to determine representativeness and other considerations for when to sample are presented below.

Industrial applicants must generally collect two types of storm water samples:

1. Grab samples collected during the first 30 minutes of discharge; and

[70] Only the part of the population served by municipal separate storm sewers is to be included in the 100,000 count, not the part served by combined sewers.

2. Flow-weighted composite samples collected during the first 3 hours of discharge (or the entire discharge, if it is less than 3 hours).

Information from both types of samples is critical to fully evaluate the types and concentrations of pollutants present in the storm water discharge.

The grab samples taken during the first 30 minutes of a storm event will generally contain higher concentrations of pollutants, since they pick up pollutants that have accumulated on drainage surfaces since the last storm event.

Composite samples characterize the average quality of the entire storm water discharge. Flowweighted composite samples provide for the most accurate determination of mass load. The flowweighted composite sample must be taken for either the first 3 hours or for the entire discharge (if the event is less than 3 hours long). Additional information on how to collect grab and composite samples is presented in Section 4.4.

Industrial applicants are required at a minimum to sample only one storm event. However, if samples from more than one storm are analyzed and the results are representative of the discharge, the data representing each event must be reported. The facility must provide a description of each storm event tested. The average of all values within the last year must be determined and the concentration, mass, and total number of storm events sampled must be reported on Form 2F. Furthermore, sampling should be conducted during normal operating procedures (day or night) and not when the facility has been closed for a period of time.

Industrial applicants must certify, as a separate requirement, that all outfalls have been tested or evaluated to determine whether nonstorm water discharges are present (e.g., process wastewater, sanitary wastes, cooling water, or rinse water) or whether illegal/illicit connections are occurring in the system. This testing should be conducted during dry weather to avoid any flows of storm water through the conveyance.

A checklist that can be used to conduct dry weather evaluations is provided in Figure 4-4.

Figure 4-4
Checklist for Conducting Dry Weather Evaluations

1. Date of inspection:____________________ 2. Facility name and address:______________
3. Date of last rain event:________________ ______________________________
4. Inspector name:____________________ ______________________________
5. Type of outfall

☐ Concrete ☐ Pipe ☐ Grassed ☐ Rock ☐ Other ______________

6. Is there visible flow from the pipe? ☐ Yes ☐ No
If yes, check all that apply. If no, go to number 7.

☐ Colored water (describe)______________ ☐ Oily sheen
☐ Odor* (describe)__________________ ☐ Sludge present
☐ Murky ☐ Clear water
☐ Floating objects (describe)____________ ☐ Stains on conveyance
☐ Absence of plant life surrounding conveyance ☐ Notable difference in plant life surrounding conveyance
☐ Scum ☐ Suds ☐ Other:______________

*e.g., rotten eggs, earthy, chemical, chlorine, soap, putrescence, gasoline, musty, etc.

Estimate the flow either visually or by describing the width, height, and shape of the conveyance and the approximate percentage of the conveyance where flow is present or the approximate depth of the flow. Describe your estimate.

7. Is there standing water present? ☐ Yes ☐ No
If yes, check all that apply. If no, go to number 8.

☐ Colored water (describe)______________ ☐ Oily sheen
☐ Odor* (describe)__________________ ☐ Sludge present
☐ Murky ☐ Clear water
☐ Floating objects (describe)____________ ☐ Stains on conveyance
☐ Absence of plant life surrounding conveyance ☐ Notable difference in plant life surrounding conveyance
☐ Suds ☐ Scum ☐ Other:______________
☐ Absence of plant life surrounding conveyance

*e.g., rotten eggs, earthy, chemical, chlorine, soap, putrescence, gasoline, musty, etc.

8. From the inspection locations, can you see any unusual piping or ditches that drain to the storm water conveyance? ☐ Yes ☐ No
9. Is there any overland flow visible from the discharge location? ☐ Yes ☐ No
10. Are there dead animals present? ☐ Yes ☐ No

Signature:

A narrative description of the method used to conduct dry weather evaluations and the date and the drainage points must be included in Section V.A of Form 2F. This statement must be certified by the appropriate party as described in Section 4.1.9.

A dry weather visual inspection is the simplest way to screen for illicit discharges. If one or more of the items on the checklist in Figure 4-4 are answered affirmatively, or if there are other reasons to believe that illicit connections exist, more detailed investigations (such as dye tests, smoke tests, evaluation of piping designs, and TV line monitoring) may be necessary.[71]

A problem with the dry weather evaluation process is that the presence of a dry weather/nonstorm water discharge may be caused by infiltration of ground or surface waters through cracks in the storm water drainage system. In this situation, all other possible sources of the nonstorm water discharge should be examined and ruled out. If no sources are found, the physical structure of the conveyance system should be inspected for deterioration.

The applicant should make every attempt to halt nonstorm water discharges to the storm sewer system unless the discharge is covered by an NPDES permit. If it is not feasible to halt the discharge of nonstorm water to the storm sewer system, and the discharge is not authorized by a process wastewater or storm water permit, the applicant must submit either Form 2C (for a process water discharge) or Form 2E (for a nonprocess water discharge) and check with state officials to see if alternate forms are required.

Industrial applicants submitting individual applications must collect and analyze a grab sample taken within the first 30 minutes of the storm event and flow-weighted composite samples from each of the industrial storm water "point source" outfalls identified on the site drainage map submitted for Section III of Form 2F. Applicants submitting quantitative data for Part 2 of the group application must also collect samples for each outfall discharging storm water associated with industrial activity. All outfalls should be sampled during the same representative storm event if possible. If this is not feasible, outfalls may be sampled during different representative storm

[71]See Sections 4.2.5 through 4.2.7.

events upon approval by the permitting authority. Descriptions of each storm event and which outfalls were sampled during each event must be included in the application. Storm water runoff from employee parking lots, administration buildings, and landscaped areas that is not mixed with storm water associated with industrial activity, or storm water discharges to municipal sanitary sewers, do not need to be sampled.

If an applicant has two or more outfalls with "substantially identical effluents," the facility may petition the permitting authority to sample and analyze only one of the identical outfalls and submit the results as representative of the other. "Substantially identical effluents" are defined as discharges from drainage areas undergoing similar activities where the discharges are expected to be of similar quantity and quality, and indistinguishable in expected composition.

4.1.6 When Municipal Facilities Must Sample

Municipal applicants are required to conduct sampling for both Parts 1 and 2 of their applications. In Part 1, municipalities must conduct a field screening analysis to detect illicit connections and illegal dumping into their storm sewer system. Where flow is observed during dry weather, two grab samples must be collected during a 24-hour period with a minimum of 4 hours between samples. These samples must be analyzed for pH, total chlorine, total copper, total phenol, and detergents (surfactants). Note that these are dry weather samples, rather than storm water samples. EPA's Guidance Manual for the Preparation of Part 1 of the NPDES Permit Applications for Discharges from Municipal Separate Storm Sewer Systems presents a description of conducting field screening sampling and provides a data sheet.

For Part 2 of the application, municipalities must submit grab (for certain pollutants) and flowweighted sampling data from selected sites (5 to 10 outfalls) for 3 representative storm events at least 1 month apart. The flow-weighted composite sample must be taken for either the entire discharge or the first 3 hours (if the event lasts longer than 3 hours). Municipal facilities are not required to collect grab samples within the first 30 minutes of a storm event.

In addition to submitting quantitative data for the application, municipalities must also develop programs for future sampling activities that specify sampling locations, frequency, pollutants to be analyzed, and sampling equipment. Where necessary (as determined by the municipality or if required by the permitting authority), responsibilities may also include monitoring industries connected to the municipality's storm sewers for compliance with their facility-specific NPDES permits. Refer to EPA's Guidance Manual for the Preparation of Part 1 of the NPDES Permit Applications for Discharges from Municipal Separate Storm Sewer Systems for information on how to develop municipal sampling programs.

4.1.7 Storm Event Criteria

Storm water discharge permit application requirements establish specific criteria for the type of storm event that must be sampled:

* The depth of the storm must be greater than 0.1 inch accumulation

* The storm must be preceded by at least 72 hours of dry weather

* Where feasible, the depth of rain and duration of the event should not vary by more than 50% from the average depth and duration.

These criteria were established to:

1. Ensure that adequate flow would be discharged;

2. Allow some build-up of pollutants during the dry weather intervals; and

3. Ensure that the storm would be "representative," (i.e., typical for the area in terms of intensity, depth, and duration).

Collection of samples during a storm event meeting these criteria ensures that the resulting data will accurately portray the most common conditions for each site. However, the permitting authority is authorized to approve modifications of this definition (especially for applicants in arid areas where there are few representative events).

In determining whether a storm is representative, there are two important steps to take. First, data on local weather patterns should be collected and analyzed to determine the range of representative storms for a particular area. Second, these results should be compared to measurements of duration, intensity, and depth to ensure that the storm to be sampled fits the representativeness criteria.

4.1.8 Obtaining Rainfall Data

Several sources provide accurate local weather information for both: (1) determining what a representative storm event is for a particular area and (2) assessing expected storm events to determine whether a predicted rainfall will be "representative," and thus, meet the requirements for storm water sampling. The National Oceanic and Atmospheric Administration (NOAA) National Climatic Data Center's (NCDC's) Climate Services Branch is responsible for collecting precipitation data. Data on hourly, daily, and monthly precipitation for each measuring station (with latitude and longitude) are available to the public on computer diskette, microfiche, or hard copy.[72]

The National Weather Service (NWS) of NOAA can also provide information on historic, current, and future weather conditions. Local NWS telephone numbers can be obtained from the NWS Public Affairs Office at (301) 713-0622. Telephone numbers are also usually in local phone directory listings under "National Weather Service" or "Weather."

4.1.9 Logistical Problems with When to Sample

Applicants may encounter weather conditions that may not meet minimum "representative" storm criteria; these conditions

[72] Orders can be placed by calling (704) 259-0682, by fax at (704) 259-0876, or by writing to NCDC, Climate Services Branch, The Federal Building, Asheville, North Carolina 28071-2733.

may prevent adequate collection of storm water samples prior to application submission deadlines. For instance, sampling may be problematic in parts of the country that experience drought or near-drought conditions or areas that are under adverse weather conditions such as freezing and flooding. Events with false starts and events with stop-start rains can also cause problems. Solutions for sampling under these circumstances are discussed below.

Where the timing of storm event sampling poses a problem, it may be appropriate for the applicant to petition the permitting authority for a sampling protocol/procedure modification either prior to sampling or after sampling is conducted (if the storm event is not acceptable). When the applicant requests a sampling protocol/procedure modification, a narrative justification should be attached. This justification should be certified by a corporate official (for industrial facilities) or the principle executive officer or ranking official (for municipalities), as per 40 C.F.R. 122.22.

Figure 4-5 summarizes common logistical problems of storm water sampling and presents solutions to the problems identified.

Figure 4-5
Common Logistical Problems of Storm Water Sampling

Problem #1:

Arid/drought areas

Solution:

Submit a petition requesting a modification to the protocol if problems are anticipated and, if it is approved, submit the application without sampling data by the application due date with a certified explanation. Provide sampling data to the permitting authority as soon as possible.

Problem #2:

Adverse weather conditions, such as freezing, flooding, winds, tornadoes, electrical storms, and gully washes

Solution:

Sample another, less hazardous event or submit a certified justification of why the event was not sampled. Provide sampling data to the permitting authority as soon as possible.

Problem #3:

False starts

Solution:

Discard the sample if the volume is inadequate. If the volume is adequate, submit the sampling data with a certified explanation that the sample is from a nonrepresentative event. Continue to monitor weather conditions and attempt to resample as soon as possible.

Problem #4:

Stop/start rains

Solution:

Continue to sample in case the storm event turns out to be representative and adequate sample volumes are obtained. If sample volumes are inadequate, continue to monitor weather conditions and attempt to resample as soon as possible.

4.1.10 Arid Areas

For arid or drought-stricken areas where a storm event does not occur prior to the time the applicant must sample and submit data with the application form, the applicant should submit the application, complete to the extent possible, with a detailed explanation of why sampling data are not provided and an appraisal of when sampling will be conducted. This explanation must be certified by the appropriate party (as described above). The applicant should also contact the permitting authority for further direction. Where the applicant can anticipate such problems, approval for an extension to submit sampling data should be acquired prior to the deadline.

4.1.11 Adverse Weather Conditions

The applicant should never conduct storm water sampling during unsafe conditions. It is likely that, in areas that experience flooding, lightning storms, high winds, etc., another representative storm event will occur for which sampling conditions will be much safer. If no other storm event occurs, the applicant should submit a justification as to why the event was not sampled. This information should be certified by the appropriate official.

4.1.12 False Starts and Stop/Start Rains

False start and stop/start rains can also cause problems. False starts may occur when weather conditions are unpredictable and it appears that a storm event may be representative, collection begins, and then the rain stops before an adequate sample volume is obtained. Some latitude may be given for the 0.1-inch rainfall requirement as long as the sample volume is adequate; the permitting authority may accept the results with applicant justification and certification.

During stop/start rains (those in which rainfall is intermittent), samples should be taken until an adequate sample volume is obtained.

4.1.13 Use of Historical Data

Data from storm water samples analyzed in the past can be submitted with applications in lieu of new sampling data if:

1. All data requirements in Form 2F are met.

2. Sampling was performed no longer than 3 years prior to submission of the permit application.

3. All water quality data are representative of the present discharge.

The historical data may be unacceptable if there have been significant changes since the time of that storm event in production level, raw materials, processes, or final products. Significant changes which may also impact storm water runoff include construction or installation of treatment or sedimentation/erosion control devices, buildings, roadways, or parking lots. Applicants should assess any such changes to determine whether they altered storm water runoff since the time of the storm event chosen for use in the permit application. Historical data can be used only in applications. Historical data cannot be used for fulfilling permit requirements.

4.1.14 Where to Take Samples

Storm water samples should be taken at a storm water point source. A "point source" is defined as any discernible, confined, and discrete conveyance, including (but not limited to) any pipe, ditch, channel, tunnel, conduit, well, discrete fissure, container, rolling stock, concentrated animal feeding operation, landfill leachate collection system, vessel, or other floating craft from which pollutants are or may be discharged.[73] Included in the definition of storm water "point sources" is storm water from an industrial facility that enters, and is discharged through, a municipal separate storm sewer. In short, most storm water discharges can be defined as "point source" discharges, since they ultimately flow into some kind of conveyance (e.g., a channel or swale).

The ideal sampling location would be the lowest point in the drainage area where a conveyance discharges storm water to U.S. waters or to an MS4. A sample point also should be easily accessible on foot in a location that will not cause hazardous sampling conditions. Ideally, the sampling site should be on the applicant's property or within the municipality's easement; if not, the field personnel should obtain permission from the owner of the property where the discharge outfall is located. Typical sampling locations may include the discharge at the end of a pipe, a ditch, or a channel.

However, logistical problems with sample locations may arise (e.g., nonpoint discharges, inaccessibility of discharge point, etc.). Logistical problems with sample locations and suggested solutions are described in Figure 4-6. In many cases, it may be necessary to locate a sampling point further upstream of the discharge point (e.g., in a manhole or inlet). If the storm water at a selected location is not representative of a facility's total runoff, the facility may have to sample at several locations to best characterize the total runoff from the site. In situations where discharge points are difficult to sample for various

---

[73] See 40 C.F.R. 122.2.

reasons, the applicant should take the best sample possible and explain the conditions in the application.

Figure 4-6
Solutions to Sample Location Problems

Problem #1:
Sampling where storm water commingles with process or nonprocess water

Solution:
Attempt to sample the storm water discharge before it mixes with the nonstorm water discharge. If this is impossible, sample the discharge both during dry and wet weather and present both sets of data to the permitting authority. This will provide an indication of the contribution of pollutants from each source.

Problem #2:
Numerous small point discharges

Solution:
Impound channel or join together flow by building a weir or digging a ditch to collect discharge at a low point for sampling purposes. This artificial collection point should be lined with plastic to prevent infiltration and/or high levels of sediment. Or, sample at several locations to represent total site runoff.

Problem #3:
Inaccessible discharge point (e.g., underwater discharges or unreachable discharges

Solution:
Go up the pipe to sample (i.e., to the nearest manhole or inspection point). If these are not available, tap into the pipe or sample at several locations to best represent total site runoff.

Problem #4:

Managing multiple sampling sites to collect grab samples during the first 30 minutes (industrial facilities only)

Solution:

Have a sampling crew ready for mobilization when forecasts indicate that a representative storm will occur or sample several different representative events. Also, for most parameters, automatic samplers may be used to collect samples within the first 30 minutes triggered by the amount of rainfall, the depth of flow, flow volume, or time.

Problem #5:

Commingling of parking lot runoff with discharge associated with industrial activity

Solution:

The combined runoff must be sampled at the discharge point as near as possible to the receiving water or the parking lot drain inlet if there is one.

Problem #6:

Sampling in manholes

Solution:

Sample in manholes only when necessary. Sampling in manholes requires training on confined space entry.

Problem #7:

Runon from other property

Solution:

If possible, estimate the volume of off-site runon contributions and off-site runon sources of pollutants to perform a mass balance calculation. Include this information in the permit application. If this estimation is not possible, provide a narrative discussion of the upstream site (e.g., is it developed; if so, the type of facility, the types of pollutants that may be present on the site, etc.).

4.1.15 Staffing Considerations

Staffing needs for sampling must be determined by the applicant. Factors in making the determination include the number of sample locations, the size of the area to be sampled, how far apart the locations are, the type of sampling required, the technique to be used, the number of samples to be taken (depending on how many parameters must be analyzed), and safety considerations.

Training sampling personnel is important to the success of storm water discharge characterization. Sampling conducted by untrained personnel may result in data that are unrepresentative of the facility's storm water discharge. These data might be rejected by the permitting authority, who would then require another sampling effort.

4.2 IDENTIFICATION OF OUTFALLS TO BE MONITORED

Form 2F requires that applicants provide quantitative data for samples of storm water discharges associated with industrial activity. If a facility discharges storm water associated with industrial activity to a municipal separate storm sewer, then the facility should sample the storm water from the site prior to discharging to the municipal separate storm sewer. Storm runoff from areas located on plant lands separate from the plant's industrial activities, such as administrative building roofs and accompanying parking lots, are not defined as storm water associated with industrial activity and hence do not need to be monitored unless the runoff is combined with storm water associated with industrial activity. 40 C.F.R. 122.21(g)(7) provides that when an applicant has two or more outfalls with substantially identical effluents, the Director may allow the applicant to test only one outfall and report that the quantitative data also apply to substantially identical outfalls.

4.2.1 Site Drainage Map

Section III of Form 2F requires that a site drainage map be attached to the application. The site drainage map must show either topography or a delineation of the drainage area served by each outfall which discharges storm water associated with industrial activity if a topographic base map is not used. The delineation of the drainage area for each outfall that discharges storm water associated with industrial activity can be based on site observations which identify drainage patterns. Drainage patterns should be shown on the site drainage map so that runoff from each drainage area drains to a separate outfall.

The site drainage map must show the location (and size - approximate for earthen structures) of all drainage conveyances or natural channels that convey or drain storm water off the applicant's property. The map must indicate whether the drainage system receiving the discharge is a natural water body, part of a municipal or nonmunicipal drainage system, or other system as applicable.

The following information must be provided and recorded on the map where appropriate:

1. Paved areas and buildings at the facility.

2. Past and present outdoor areas used for storage or disposal of significant materials.

3. Hazardous waste treatment, storage or disposal facilities, or accumulation areas (including those not requiring a RCRA permit).

4. Injection wells.

5. Material loading and access areas (e.g., loading docks and main truck routes on the facility property).

6. Areas where pesticides, herbicides, soil conditioners, and fertilizers are applied.

7. Structural control measures to reduce pollutants in storm water runoff.

8. Surface water bodies which receive storm water discharges from the facility.

During the preparation of a site drainage map, or the review of an existing one, emphasis should be placed on the identification of all inflow sources to ensure that inappropriate sources of nonstorm water entry are not present. The map should identify points of entry to the facility site storm water drain system, including catch basins, floor drains, and roof leaders.

The site drainage map required in Form 2F should show the location and an identifying number or name for each storm water outfall at the facility.

4.2.2 Evaluation of Presence of Nonstorm Water Discharges

Form 2F requires applicants to certify that all outfalls that discharge storm water associated with industrial activity have been tested or evaluated for the presence of nonstorm water discharges. Applicants do not have to test for the presence of nonstorm water discharges already subject to a NPDES permit. Acceptable procedures include: dry weather observations of outfalls or other appropriate observation locations; the analysis and validation of accurate piping schematics; dye tests; or other procedures for ensuring that there are no inappropriate connections or discharges to the storm drain system. The permit application requires a description of the method used, the date of testing (if applicable), and the onsite drainage locations observed during the test. Any nonstorm water discharge detected which is not already identified in a NPDES permit must be identified in Form 2C (for process wastewater) or Form 2E (for non-process wastewater) which must accompany the storm water discharge application (Form 1 and Form 2F).

The following subsections provide a description of several procedures that can be used in developing a certification and an overview of the applicability of the tests and the resources required for performing the tests. A first step should be to identify potential sources of nonstorm water at the facility and to focus on those places.

4.2.3 Visual Inspection of Storm Drain

A visual inspection of the system conducted during dry weather can be an effective method of locating illicit connections to the storm drain system. The observation should be made during normal business hours when sources of nonstorm water are typically operating. A record should be kept of all observed flows and any stains, sludges, or other abnormal conditions observed. Where flows are observed, additional analysis, such as dye testing (described below) may be necessary to identify the source of the flows.

This method is applicable to any industrial site with a storm drain system where an outfall or other location (e.g. manhole) down gradient from potential nonstorm water discharges can be observed.No special equipment is required.

4.2.4 Review and Validation of Piping Schematics Description

A careful review of piping schematic drawings for industrial sites can identify the intended routing of flows from particular areas or drains. This review should be accompanied by visual inspection to compare the "as built" condition to the plans and to determine whether any unrecorded piping modifications have been made.

This method is most applicable for industrial sites which have large or elaborate piping arrangements, usually recorded on schematic piping drawings. It is most applicable in conjunction with use of the other techniques described below. No special equipment is required, though dye tests may be useful in specific situations to clarify discrepancies which cannot be resolved visually.

4.2.5 Dye Tests Description

Dye testing involves releasing fluorescent, nontoxic dye into the suspected source of nonstorm water, (e.g., a drain, sink, toilet, or pipe) and checking to see whether the dye shows up in the storm water outfall. A quantity of dye is released at the selected location while an observer watches for the dye at a downstream location. If the inlet is discharging to the storm drain, the dye will be detected at the downstream location. Dye doses should be sufficiently large so that the dye at the downstream location is visible to the naked eye.

Dye tests are most effective for determining if an identified drain or catchbasin is connected to the storm sewer where the outfall of the storm sewer is submerged, but the receiving water can be observed. (Where the outfall or other point can be observed and is not submerged, dry weather observation can be made or water can be used instead of a dye). Dye tests can also be used where dry weather flows have been observed, but the source of the flow has not yet been observed. It is best used when there are only a limited number of possible sources of nonstorm water to the storm drain that need to be investigated.

No special equipment is necessary to conduct a dye test. Dye is the only material required. Effective dyes that are safe and harmless are available in powder, tablet, or liquid form. A minimum field crew of two is needed, one to apply the dye, the other to observe the storm drain.

4.2.6 Smoke Testing

Smoke testing involves pumping smoke into a storm sewer and viewing the facility to see if smoke escapes through unknown openings or storm sewer inlets. The presence of smoke indicates that storm water may enter the sewer through these openings or inlets. However, smoke testing may prove ineffective at finding nonstorm water discharges to separate storm sewers. Smoke passage may be blocked due to line traps that are intended to block sewer gas.

4.2.7 TV Line Surveys Description

TV line monitoring is a technique whereby a small video camera is placed in the storm sewer and a video image of the sewer is viewed on a monitor at the surface to identify illicit connections. The camera can be moved through the sewer by remote control.

TV surveys are conducted with a mobile closed-circuit television system consisting of a monitor screen, camera, drag lines, and reels and cables that allow the camera to be guided through a section of pipeline. The TV picture allows a visual inspection of the interior of the drain pipe and can be used for pipelines with diameters that range from 4 inches to approximately 48 inches. Television inspection of a storm drain provides positive information (and a documented record) of the interior of the pipelines. All inlets to the line can be identified and located. Systems for conducting TV surveys can be purchased, leased, or rented. Alternatively, a firm which specializes in this work can be hired.

TV surveys may serve as useful tools where an initial survey identifies a nonstorm water discharge and the operator is having difficulty in finding the source. A TV survey can locate entry points to the storm drain system, determine whether or not there is flow in them, and permit estimates of the flow to be made. However, in many cases these observations will need to be supplemented by other methods to identify the specific source (above ground) of the connection. This may be accomplished by inspection of drain maps, dye tests, or possibly smoke tests.

Resources required for a TV survey of storm drains include (1) a TV camera; (2) a TV monitor and VCR to record the survey; and (3) a rig consisting of video cables, tow lines, and related equipment for properly guiding the camera in the line at a controlled rate, recording distance moved, and withdrawing the camera from the pipeline.

The cost to conduct a TV survey can range from $1 to $3 per foot of storm sewer. For small surveys costs could vary from $125 to $200 per hour, including labor and rental of the necessary equipment. However, this cost can increase significantly if the storm sewer must be cleaned of debris prior to conducting the TV survey. On average, approximately 1000 feet of sewer can be inspected in a day. In a clean sewer, up to 2000 feet can be inspected.[74]

4.3 ESTIMATES OF DISCHARGE FLOW RATES AND VOLUMES

Form 2F requires applicants to provide quantitative data based on samples collected during storm event(s). One set of parameters that must be provided for such storm event(s) are flow estimates or flow measurements and an estimate of the total volume of the discharge. The method of flow estimation or measurement must be described in the application. EPA intends that applicants need only provide rough estimates of flows in Form 2F. This section discusses methods for obtaining the required information.

---

[74] For more information on smoke and dye testing and TV line monitoring, consult EPA's Guidance Manual for the Preparation of NPDES Permit Applications for Storm Water Discharges Associated with Industrial Activity, EPA-505/8-91002 (April 1991).

4.3.1 Estimating Flows and Volumes

Runoff flow rates and volumes can be estimated by using the total rainfall amount for the storm event and estimated runoff coefficients for the facility. Runoff coefficients represent the fraction of total rainfall that will be transmitted as runoff from the facility. As such, the coefficients reflect the ground surface or cover material. To estimate runoff volume and rates, it can be assumed that paved areas and other impervious structures such as roofs have a runoff coefficient of 0.90 and, therefore, 90% of the rainfall is conveyed from the facility as runoff. For unpaved surfaces, it can be assumed that the runoff coefficient is about 0.50. The total volume of discharge for the event is then estimated by:

total runoff volume (cubic ft) = total rainfall (ft) x [facility paved area x 0.90 + facility unpaved area x 0.50]

The facility areas used in this calculation should be in units of square feet and should include only those areas drained by the outfall sampled. To estimate an average flow rate, divide the volume by the duration of the rainfall event. If desired, a more accurate estimate can be made by using more specific runoff coefficients for different parts of the facility based on the type of ground cover.

There are a variety of techniques for measuring or estimating flow rates. Flow measuring devices based on pipe invert sections (e.g., flumes, weirs, and others) are commercially available. For locations that may be used for routine monitoring in the future, the applicant may consider installing these types of devices for ease in future measurements. The installed cost of a weir, for example, typically ranges from about $1,000 to $5,000. Once installed, the weir must be calibrated so that future measurements of stage (i.e., depth of flow) can be converted directly to flow volumes. The installation and calibration of such devices should be performed by experienced personnel.

To estimate flow rates in units of volume per time, such as cubic feet per second, information on flow velocities and depth of flow are required. Flow rate estimates may be obtained by measuring depth of flow and velocity in a pipe of known diameter or other conveyance structure at frequent intervals during a storm runoff event. For a pipe or other structure of known size, the cross-sectional area of flow can be calculated for any depth of flow using geometric relationships. Flow velocities can be measured by using suitable units (e.g., propeller operated devices) attached to a portable current meter. Flow velocity measurements should be obtained from representative locations throughout the flow cross-section. Such units are commercially available at costs ranging from about $1,000 to $3,000. While these devices may be fairly expensive, they are easy to use and they provide accurate data if used properly.

Flow velocities can be estimated using simpler methods, such as measuring the time of passage of an object (e.g., an orange) between two points a known distance apart (e.g., manholes).

4.3.2 Estimation of Flow Rates Based on Flow Velocity

If the measurements of flow depth are recorded and converted to crosssectional areas (in square feet), and the corresponding velocities for each depth are recorded (in feet per minute), then the flow rate (Q) in cubic feet per minute (cfm) is:

$$Q = (\text{area})(\text{velocity})$$

The maximum flow rate is the highest value recorded during the storm event. The time-weighted average flow rate for the storm event can be estimated by the average of the individual values recorded.

4.3.3 Estimation of Volumes Based on Flow Rate Estimates

The total volume of discharge can be estimated by first multiplying each of the flow rates determined above by a time interval that represents the portion of the total storm duration associated with the measurement, and then adding all such partial volumes. If the time intervals used are seconds, then the total flow of runoff will be in units of cubic feet.

A procedure for calculating the total runoff volume from a set of discrete measurements of flow depth and velocity during a storm runoff event are as follows:

Step 1: Measure and tabulate flow depths and velocities every 20 minutes during at least the first 3 hours of the runoff event.

Step 2: Calculate and tabulate the cross-sectional area of flow for each of the flow depths measured. Calculate the flow rate (Q) for each discrete set of flow rate and flow velocity measurements. Q = (area)(velocity).

Step 3: Plot flow rate, Q versus time.

Step 4: Assign each flow rate measurement a duration equal to the sum of $\frac{1}{2}$ the time interval between the preceding and succeeding measurements. In the ideal case of uniform 20 minute intervals, the durations are [$(20)\frac{1}{2} + (20)\frac{1}{2} = 20$ minutes].

Step 5: Compute the flow volume associated with each observation (V_1, V_2, ..., V_9) by multiplying the measured flow rate by the duration (in this case, 20 minutes). Be sure the units are consistent. For example, if durations are in minutes and flow velocities are in cubic feet per second (cfs), convert the durations to seconds or the velocities to feet per minute.

Volume (V) = Flow Rate (cfm) x Duration (minutes)

Step 6: The beginning volume can be approximated by assuming that the flow rate is zero at time zero and increases linearly to the first calculated flow rate (Q_1) at 20 minutes.

The final volume can be approximated similarly by assuming that flow drops uniformly from the last calculated flow rate (Q_9) to zero at the time when Q_{10} would have been taken.

Step 7: Total the individual volumes calculated in Step 5 with the initial and final volume approximations calculated in Step 6 to obtain the total runoff volume.

4.4 COLLECTING STORM WATER DISCHARGE SAMPLES

This section provides guidance for collecting grab samples, flow-weighted composite samples, and identifying the constituents or parameters that must be monitored. Because of the variable nature of storm water flows during a rainfall event and different analytical considerations for certain pollutants, the storm water regulations establish specific requirements for sample collection techniques. The quality of storm water discharges and logistical needs for sampling will be different for industrial applicants and municipal applicants. Therefore, specific sampling requirements vary. The applicant should carefully plan his/her sampling strategy prior to the actual sampling event, e.g., walk the site to determine appropriate sampling locations, become familiarized with local rainfall patterns, train sampling staff in procedures and safety, consult with laboratory, and collect supplies.

4.4.1 Sample Type and Technique

There are three basic aspects of sampling:

* Sample type (i.e., grab versus composite).

* Sample technique (i.e., manual versus automatic).

* Flow measurement methods.

It is important to understand the difference between sample type and technique. "Sample type" refers to the kind of sample that must be collected - either a grab or a composite sample. "Sample technique" refers to the method by which a grab or composite sample is actually collected - either manually or by automatic sampler.

To comply with storm water application requirements, the sample type (grab or composite) must be collected in accordance with 40 C.F.R. 122.21(g)(7) and 40 C.F.R. Part 136. The storm water application requirements clearly specify which pollutants must be analyzed by grab sample and which by composite sample. Although the requirements in 40 C.F.R. 122.21(g)(7) do not explicitly specify either manual or automatic sampling techniques, the approved analytical methods contained in 40 C.F.R. Part 136 direct that grab samples must be collected manually for certain pollutants.

4.4.2 Grab Samples

A grab sample is a discrete, individual sample taken within a short period of time (usually less than 15 minutes). Analysis of grab samples characterizes the quality of a storm water discharge at a given time of the discharge.

A grab sample must be collected during the first 30 minutes of the runoff (or as soon thereafter as practicable). The sample collected should be large enough for all of the laboratory analyses to be performed, but at least 100 milliliters (ml). Grab samples are typically collected by filling the sample container just below the water surface in the flow channel. Extension rods or cables can be used to reach inaccessible locations. The grab sample should be collected from near the center of the flow channel, where turbulence is at a maximum (and therefore the storm runoff is well mixed), or at a site specified in an existing permit, or at any site adequate for the collection of a sample that would be representative of the storm water quality.

All samples must be properly handled (i.e., holding time prior to analysis, storage temperature, preservation methods) and analyzed by the methods contained in 40 C.F.R. Part 136. Most commercial laboratories will be familiar with these requirements and can provide information on appropriate handling procedures. Quality assurance/quality control (QA/QC) methods must be implemented both in the field by the applicant and in the lab to ensure the accuracy and validity of the analytical results. Most labs can assist applicants in designing a field QA/QC program and will also provide sample containers that are suitable (e.g., container material, type, and size) to the analysis to be performed. The labs will also typically report to the applicant the results of their internal QA/QC upon request.

If an analytical method is not listed in 40 C.F.R. 136 for a particular pollutant, then the applicant may use any suitable method for measuring the level of the pollutant in the discharge provided that the applicant submits a description of the method or a reference to a published method. The description should include the sample holding time, preservation methods, and the quality control measures used.

The parameters pH and temperature are time dependent and must be measured in the field at the time of sample collection rather than in the laboratory.

4.4.3 Flow-Weighted Composite Samples

A composite sample is a mixed or combined sample that is formed by combining a series of individual and discrete samples of specific volumes at specified intervals. Although these intervals can be time weighted or flow weighted, the storm water regulations require the collection of flowweighted composite samples. This means that discrete aliquots, or samples, are collected and combined in proportion to flow rather than time. Composite samples characterize the quality of a storm water discharge over a longer period of time, such as the duration of a storm event.

A flow-weighted composite sample is a single sample intended to provide the average water quality for the entire runoff event. Because this type of sample accounts for variations in flow that occur during an event, water quality data from a flow-weighted composite sample is considered to be more representative of the average runoff quality for other methods such as a time-weighted composite.

A flow-weighted composite sample can be collected during either the entire runoff event (which may be less than 3 hours) or during at least the first 3 hours of the runoff. The sample can be collected using either automatic sampling equipment or by manually collecting and combining a series of discrete grab samples (aliquots) in an appropriate manner. In either case, appropriate procedures must be followed to obtain a sample for analysis that is flow weighted and, hence, will provide an indication of the average (or event mean) concentration for the storm runoff event.

4.4.4 Application Requirements

Both types of samples must be collected and analyzed for storm water discharge permit applications.

Grab samples must be collected for the following conditions:

1. For storm water discharges associated with industrial activity, a grab sample must be obtained during the first 30 minutes of a discharge. This requirement is in addition to the composite sampling requirements. These samples are intended to characterize the maximum concentration of a pollutant that may occur in the discharge and/or may indicate commingling of nonstorm water discharges.

2. For storm water discharges from large and medium municipal separate storm sewers, grab samples are required for Part 1 of the application if a discharge is noted during dry weather field screening. Two grab samples must be collected during a 24-hour period with a minimum of 4 hours between samples. These samples are intended to assist in the identification of illicit connections or illegal dumping. In Part 2, grab samples may be required for the analysis of certain pollutants for which municipalities are required to sample.

Flow-weighted composite samples must be collected during the first 3 hours of discharge or the entire discharge (if it is less than 3 hours) for both industrial and municipal applicants.

The regulations at 40 C.F.R. 122.21(g)(7) identify certain pollutants for which grab sampling is required. Monitoring by grab sample must be conducted for pH, temperature, cyanide, total phenols, residual chlorine, oil and grease (O&G), fecal coliform, and fecal streptococcus. Composite samples are not appropriate for these parameters due to their tendency to transform to different substances or change in concentration after a short period of time. Such transformations may be particularly likely in the presence of other reactive pollutants.

4.4.5 Manually Collected Samples

A manually collected composite flowweighted sample can be prepared by the following procedures. Collect samples of the same size (at least 100 ml and preferably 1000 ml) at regular intervals during the duration of the entire runoff event or for at least the first 3 hours of the event. Samples should be collected every 20 minutes to meet the requirement of at least 15 minutes between sample collection times. Storm runoff flow rates and flow cross-sectional areas in the conveyance should be estimated each time an individual sample is taken. Relative flow rates rather than actual flow rates can be used. Where flow rates are estimated based on runoff coefficients, then the amount of rainfall during a given time period should be measured or estimated, and discharge flow rates assumed to be proportional to the amount of rainfall occurring during a given interval. Remove a portion (or aliquot) from each of the individual samples that is proportional to the flow rate for that time interval (there should be at least nine individual samples, i.e., three samples collected each hour during the first 3 hours of runoff) and combine them in the container that will be sent to the laboratory for analysis. Only the composite sample needs to be sent to the laboratory for analysis. The actual amount taken from each of the individual samples should be in proportion to the flow rate or volume of flow associated with that sample.

4.4.6 Automatic Samplers

Automatic samplers are labor-saving devices, but are fairly expensive to purchase. The samplers consist of an intake device set in the channel which is attached by tubing to a pump that can draw a sample from the storm drain into a sample bottle. However, in order for the sample obtained to be a flow-weighted composite, the automatic sampler must be triggered by the flow-sensing device. Samples of fixed volume are collected each time the flow-sensing device indicates that a specified quantity of flow has passed the sample point.

An appropriate flow-sensing device, coupled to the automatic sampler, is necessary for the automatic system to produce a flow-weighted composite. If the monitoring equipment does not employ such a coupled system, then the automatic sampler merely serves as a mechanical means for withdrawing the sample (usually at fixed time intervals).

Automatic samplers generally range in price from about $8,000 to $16,000 for equipment costs alone. Units with telemetry are in the upper end of this range. The equipment included with a standard unit includes a fabricated weir, an automatic sampler with silica sample containers, software to control the remote computer data logger, housing for unit, thermistor, and pressure sensor. The installation and flow rating of a unit will cost approximately $6,000 to $8,000, depending on whether the unit is installed in a manhole, open culvert or channel, or stream. Digital doppler velocity sensors can also be purchased and installed. Such units would replace the weir, data logger, and pressure sensor identified above.

4.4.7 Representative Storm Event

Section VII of Form 2F requires that specific pollutants in storm water discharges be measured and reported as concentrations and as total mass. At least one representative storm event must be sampled to collect this information. If samples from more than one storm are analyzed and the results are representative of the discharge, the results must be reported in Section VII of Form 2F.

A representative storm is a storm that is "typical" for the area in terms of intensity, volume, and duration. The storm must have a volume greater than 0.1 inch, must be preceded by at least 72 hours of dry weather, and should not vary by more than 50% from the average rainfall volume and duration.

A representative storm event must be sampled to provide water quality data for the initial runoff period (i.e., a grab sample to measure first-flush effects). A flow-weighted composite sample must also be collected and analyzed separately from the grab sample to provide an estimate of the average runoff water quality for the storm event. Data from samples analyzed in the past may be used, provided that:

* All data requirements in Form 2F are met;

* Sampling was done no more than three years before submission of the permit application; and

* All water quality data are representative of the present discharge.

Among the factors which would cause the data to be unrepresentative are significant changes in production level, changes in raw materials, processes, or final products, and significant changes in storm water management activities.

Grab samples and flow-weighted composite samples must be collected and analyzed from each of the storm runoff outfalls identified on the site drainage map in Section III of Form 2F. However, if an applicant has two or more substantially identical outfalls, they may request permission from the permitting authority to sample and analyze only one outfall and submit the results of the analysis for the other substantially identical outfalls. Substantially identical outfalls are those from drainage areas undergoing similar activities where the discharges are expected to be of similar water quality. If the request is granted, identify which outfall was tested and describe why the outfalls which were not tested are substantially identical. Provide this information on a separate sheet attached to the application form.

4.4.8 Sampling at Retention Ponds

Retention ponds with greater than a 2-hour holding time for a representative storm event may be sampled by grab sample. Composite sampling is not necessary. The rationale for this is that, because the water is held for at least 24 hours, a thorough mixing occurs within the pond. Therefore, a single grab sample of the effluent from the discharge point of the pond accurately represents a composite of the storm water contained in the pond. If the pond does not thoroughly mix the discharge, thereby compositing the sample, then a regular grab and composite sample should be taken at the inflow to the pond. Since each pond may vary in its capability to "composite" a sample, applicants must carefully evaluate whether the pond is thoroughly mixing the discharge. Such factors as pond design and maintenance are important in making this evaluation. Poor pond design, for example, where the outfall and inflow points are too closely situated, may cause shortcircuiting and inadequate mixing. In addition, poor maintenance may lead to excessive resuspension of any deposited silt and sediment during heavy inflows. Because of factors such as these, the applicant should determine the best location to sample the pond (e.g., at the outfall, at the outfall structure, in the pond) to ensure that a representative composite sample is taken. If adequate compositing is not occurring within the pond, the applicant should conduct routine grab and flowweighted composite sampling.

A grab sample and a flow-weighted sample must be taken for storm water discharges collected in holding ponds with less than a 24-hour retention period. The applicant must sample the discharge in the same manner as for any storm water discharge, as described in 40 C.F.R. 122.21(g)(7).

4.4.9 Pollutants to be Analyzed

Section VII of Form 2F requires that several common pollutants must be analyzed in both the grab sample and the flow-weighted composite sample, while additional analyses are dependent upon existing NPDES permit conditions or whether the discharger has reason to believe other pollutants may be present in the storm runoff discharge. A separate table should be completed for each outfall. Remember that 40 C.F.R. 122.21(g)(7) provides that rather than using a flow-weighted sample for quantitative data for pH, temperature, cyanide, total phenols, residual chlorine, oil and grease, fecal coliform, and fecal streptococcus, a grab sample must be used.

Part A of Section VII requires that both grab samples and flow-weighted composite samples be analyzed for:

* Biochemical oxygen demand (BOD_5)
* Chemical oxygen demand (COD)
* Total suspended solids (TSS)
* Total Kjeldahl Nitrogen (TKN)
* Nitrate plus nitrite nitrogen
* Total phosphorus

In addition, grab samples must be analyzed for pH.

Part B of Section VII requires that each pollutant limited in an effluent guideline to which the facility is subject or any pollutant listed in the facility's NPDES permit for its process wastewater (if the facility is operating under an existing permit) be analyzed for and reported separately for each outfall in Part B.

Part C of Section VII requires the listing of any pollutant shown in Tables 2F-2, 2F-3, and 2F-4 that the discharger knows or has reason to believe is present in the discharge and was not already identified above.

Table 2F-2 includes conventional and nonconventional pollutants. For any pollutant from this table listed in Part C, the applicant is required to either report quantitative data or briefly describe the reason the pollutant is expected to be discharged.

Table 2F-3 lists toxic pollutants. For every pollutant listed in Table 2F-3 that is expected to be discharged in concentrations of 10 parts per billion (ppb) or greater, the applicant is required to submit quantitative data. For acrolein; acrylonitrile; 2,4 dinitrophenol; and 2-methyl-4, 6 dinitrophenol, the applicant must submit quantitative data if these four pollutants (collectively) are expected to be discharged in concentrations of 100 ppb or greater. For every other pollutant listed in Table 2F-3 that is expected to be discharged in concentrations less than 10 ppb (or 100 ppb total for the four pollutants listed above), then the applicant must either submit quantitative data or briefly describe the reasons the pollutant is expected to be discharged.

Table 2F-4 lists hazardous substances. For each outfall, the applicant must list any pollutant from Table 2F-4 that is known or believed to be present in the discharge and explain why they believe it to be present. No analysis is required, but if the applicant has analytical data, it must be reported.

Under 40 C.F.R. 117.12(a)(2), certain discharges of hazardous substances (listed in 40 C.F.R. 177.21 or 40 C.F.R. 302.4) may be exempted from the requirements of CWA Section 311, which establishes reporting requirements, civil penalties, and liability for cleanup costs for spills of oil and hazardous substances. A discharge of a particular substance may be exempted if the origin, source, and amounts of the discharged substances are identified in the NPDES permit application or in the permit, if the permit contains a requirement for treatment of the discharge, and if the treatment is in place. To apply for an exclusion of the discharge of any hazardous substance from the requirements of Section 311, attach additional sheets of paper to the form and provide for the following information:

1. The substance and the amount of each substance which may be discharged.

2. The origin and source of the discharge of the substance.

3. The treatment which is to be provided for the discharge by:

a. An onsite treatment system separate from any treatment system treating the normal discharge;

b. A treatment system designed to treat the normal discharge and which is additionally capable of treating the amount of the substance identified under paragraph 1 above, or

c. Any combination of the above.

4.4.10 Reporting

All sampling data obtained for the purpose of completing Section VII of Form 2F must be reported as concentration and as total mass. The applicant may report some or all of the required data by attaching separate sheets of paper instead of filling out pages VII-1 and VII-2 if the separate sheets contain all the required information in a format which is consistent with pages VII-1 and VII-2 in spacing and in identification of pollutants and columns. Use the following abbreviations in the columns headed "Units":

ppm = parts per million
mg/l = milligrams per liter
ppb = parts per billion
ug/l = micrograms per liter
lbs = pounds
ton = tons (English tons)
mg = milligrams
g = grams
T = tonnes (metric tons)
kg = kilograms

All reporting of values for metals must be in terms of "total recoverable metal" unless:

* An applicable promulgated effluent limitation or standard specifies the limitation for the metal in dissolved, valent, or total form.

* All approved analytical methods for the metal measure only its dissolved form (e.g., hexavalent chromium).

Chapter 5

POLLUTION PREVENTION PLAN DEVELOPMENT AND IMPLEMENTATION

5.1 INTRODUCTION

The purpose of this chapter is to provide guidance to facilities in the development of storm water pollution prevention plans. This chapter outlines the different pollution prevention plan requirements of National Pollutant Discharge Elimination System (NPDES) permits, explains various baseline Best Management Practices (BMPs), and offers a systematic approach for developing and implementing an effective pollution prevention plan.

5.2 POLLUTION PREVENTION PLAN DEVELOPMENT

Completion of the following preliminary tasks will help facilitate development of a Storm Water Pollution Prevention Plan:

* Decide who will be responsible for developing and implementing the Storm Water Pollution Prevention Plan.

* Look at other existing environmental facility plans for consistency and overlap.

By following these two basic steps, the facility can better organize its staff and lay the necessary groundwork for the Pollution Prevention Plan.

5.2.1 Pollution Prevention Team

The pollution prevention plan provisions of EPA's general permit require that the permittee choose a pollution prevention team. The regulations require that each plan identify a specific individual or individuals within the facility organization as members of a storm water Pollution Prevention Team that are responsible for developing the Storm Water Pollution Prevention Plan and assisting the facility or plant manager in its implementation, maintenance, and revision. The plan must clearly identify the responsibilities of each team member. The activities and responsibilities of the team shall address all aspects of the facility's Storm Water Pollution Prevention Plan.[75]

5.2.2 Selecting Team Members

Designating a specific individual or team to develop and implement the pollution prevention plan serves several purposes. Naming the individual or team members makes it clear that part of that person's job is to prevent storm water pollution. Identifying a specific individual also provides a point of contact for those outside the facility who may need to discuss aspects of the facility's pollution prevention plan (i.e., regulatory officials, etc.).

When setting up a pollution prevention team, it is important to identify the key people on site who are most familiar with the facility and its operations, and to provide adequate structure

---

[75] See Final NPDES General Permits for Storm Water Discharges Associated With Industrial Activity, Part IV.D.1., 57 Fed. Reg. 44438, 44447 (Sept. 25, 1992).

and direction to the facility's entire storm water management program. The pollution prevention team concept is flexible and should be molded to conform to the resources and specific conditions of the facility. Specific activities of the pollution prevention team, the number of members, and their background and experience will vary for each facility.

Effective organization of the pollution prevention team is important in order for the team to accomplish the task of developing and implementing a comprehensive Storm Water Pollution Prevention Plan. There are two important features in organizing a team of this nature: (1) selecting the right individuals to serve on the team and (2) establishing good channels of communication.

The designated individual or team will be the driving force behind the development, implementation, maintenance, and revision of the facility's Storm Water Pollution Prevention Plan. One of the first tasks of those responsible is to define and agree upon a clear and reasonable set of goals for the facility's overall storm water management program. Where a team is involved, the responsibilities or duties of specific team members should be clearly defined.

Areas of responsibilities include initial site assessment, identification of pollutant sources and risks, decision making on appropriate BMPs, directing the actual implementation of the BMPs, and then, regular evaluations to measure the effectiveness of the plan.

To ensure that the Storm Water Pollution Prevention Plan remains effective, the person or team responsible for maintaining the pollution prevention plan must be aware of any changes that are made in plant operations to determine if any changes must be made.

While a designated individual or a pollution prevention team can be assigned the job of developing and implementing a Storm Water Pollution Prevention Plan, plant management is ultimately responsible for the implementation of the plan and for compliance with all applicable storm water requirements. Accordingly, the designated individual or team must have a clear line of communication with plant management to ensure that they are able to function in a cooperative partnership.

Any team, by definition, involves decision making and planning in a group setting. This allows for people with different ideas and areas of expertise to share knowledge and collectively determine what works best for a particular facility. To broaden the base of involvement in the facility's storm water pollution prevention program, team members should represent all phases of the facility's operations.

For example, at a large facility, a team may be comprised of representatives from plant management, all aspects of production operations, engineering, waste handling and treatment (environmental department), and, if applicable, research and development.

Not all facilities will have or require all of these "team" positions. Team membership depends on the type of operations occurring at a facility. For example, a small trucking operation may find it appropriate to designate a single individual or a very small pollution prevention team with experience in key types of facility operations, such as vehicle maintenance, vehicle washing, fueling, and materials handling.

For a facility that has already designated a spill prevention and response team, the facility may use some of these personnel on the storm water pollution prevention team, thus overlapping the two groups to a certain extent. However, the roles and responsibilities of the pollution prevention team reach beyond the activities of a spill prevention and response team, and consequently, it would not be appropriate for a facility simply to substitute the spill response team for the pollution prevention team without clearly examining the roles and requirements related to storm water management.

5.2.3 Facilities Subject to EPCRA Section 313 Reporting

EPA's general permit contains more specific pollution prevention team requirements for facilities subject to EPCRA Section 313 reporting for water priority chemicals. The team must designate a person who will be accountable for spill prevention at the facility and identify this person in the plan. The designated person is responsible for setting up necessary spill emergency procedures and reporting requirements to isolate, contain, and clean up spills and emergency releases of

EPCRA Section 313 water priority chemicals before a discharge can occur.[76]

5.2.4 Consistency with Other Environmental Plans

The pollution prevention plan provisions of EPA's General Permit allow the permittee to incorporate provisions of Spill Prevention Control and Countermeasure (SPCC) plans or BMPs from other NPDES permits into Storm Water Pollution Prevention Plans. The regulations state that "Storm Water Pollution Prevention Plans may reflect requirements for Spill Prevention Control and Countermeasures (SPCC) plans developed for the facility under Section 311 of the Clean Water Act or BMP programs otherwise required by an NPDES permit for the facility as long as such requirement is incorporated into the Storm Water Pollution Prevention Plan."[77]

Many industrial facilities may have already incorporated storm water management practices into day-to-day operations as a part of an environmental management plan required by other regulations. Potentially relevant elements of a number of different types of plans may be incorporated into the Storm Water Pollution Prevention Plan, including the following plans:

* RCRA Preparedness, Prevention and Contingency Plan (40 C.F.R. Parts 264 and 265).
* SPCC requirements (40 C.F.R. 112).
* NPDES Toxic Organic Management Plan (40 C.F.R. 413, 433, 469).
* OSHA Emergency Action Plan (29 C.F.R. 1910).

It is the responsibility of the pollution prevention team to evaluate these other plans to determine which, if any, provisions may be incorporated into the Storm Water Pollution Prevention Plan.

In some cases, it may be possible to build on elements of these plans that are relevant to storm water pollution prevention. For example, if the facility already has an effective SPCC plan in place, elements of that spill prevention strategy may be

---

[76] Id., Part IV.D.7.b.(9), 57 Fed. Reg. 44438 (Sept. 25, 1992).
[77] Id., Part IV.D.6., 57 Fed. Reg. 44438 (Sept. 25, 1992).

relevant to the approach taken for storm water pollution prevention. More specifically, lists of potential pollutants or constituents of concern may provide a starting point for a list of potential storm water pollutants.

Although the facility should try to build on relevant portions of other environmental plans, it is important to remember that the facility's Storm Water Pollution Prevention Plan must be a comprehensive, stand-alone document.

5.3 POLLUTION SOURCE ASSESSMENT

After identifying who is responsible for developing and implementing the facility's Storm Water Pollution Prevention Plan, a pollutant source assessment should be performed. The assessment is used to evaluate what materials or practices are or may be a source of storm water contaminants at the facility site. To complete this process, the facility will need to:

* Assess the potential sources of storm water pollution at the facility.

* Create a map of the facility site to locate pollutant sources and determine storm water management opportunities.

* Conduct a material inventory.

* Evaluate past spills and leaks.

* Identify nonstorm water discharges and illicit connections.

* Collect or evaluate storm water quality data.

* Summarize the findings of the pollution source assessment.

The pollutant source assessment phase is designed to target the most important pollutant sources for corrective and/or preventive action, thus using a "risk-based" approach to environmental protection. Details on how to complete this assessment are provided in the subsections that follow.

5.3.1 Description of Potential Pollutant Sources

The pollution prevention plan provisions of EPA's general permit require that the permittee provide a description of potential sources of storm water pollution. Part IV.D.2. of the general permit requires that "Each plan shall provide a description of potential sources which may reasonably be expected to add significant amounts of pollutants to storm water discharges or which may result in the discharge of pollutants during dry weather from separate storm sewers draining the facility. Each plan shall identify all activities and significant materials which may potentially be significant pollutant sources."[78]

5.3.2 Drainage and Site Map

In order to complete the description of potential pollutant sources, the permittee must develop a facility site map showing storm water outfalls and site drainage areas.[79] The facility site map must include:

* An outline of the drainage area of each storm water outfall.

* Location of any existing structural control measures used to reduce pollutants in storm water runoff.

* Surface water bodies.

* Locations where significant materials are exposed to precipitation.

* Locations where major spills or leaks have occurred.

* Locations for each of the following activities (where exposed to storm water):

[78] Id., Part IV.D.2., 57 Fed. Reg. 44438, 44447 (Sept. 25, 1992).
[79] Id., Part IV.D.2.a.(1), 57 Fed. Reg. 44438, 44447 (Sept. 25, 1992).

- Fueling stations
- Vehicle and equipment maintenance and/or cleaning areas
- Loading/unloading areas
- Treatment, storage, or waste disposal areas
- Liquid storage tanks
- Processing areas
- Storage areas.

The facility site map is basically an illustration of the overall site and location, and should indicate property boundaries, buildings, and operation or process areas, as well as provide information on drainage, storm water control structures, and receiving streams. Locating these features on the map will help to assess where potential storm water pollutants are located, where they mix with storm water, and where storm water leaves the site. All of this information is essential in identifying the best opportunities for storm water pollution prevention or control.

Outfalls and Drainage Areas. Once boundaries and facility structures have been shown on the site map, identify all of the storm water outfalls (also called "discharge points") on the facility site. A storm water outfall is the point where storm water enters a natural waterway or a separate storm sewer system. If the facility has its own storm water conveyance system, locate where the pipes or conveyances discharge to a stream, river, lake, or other water body. If the facility discharges to a municipal separate storm sewer system, the on-site drainage point into the system is an outfall. However, on many sites, storm water is simply collected in ditches. The discharge points may not be so obvious, especially when it is not raining. In these cases, it may be necessary to inspect the site during a rain storm to identify discharge points. Clearly label each outfall either with letters (A, B, C, etc.) or numbers (1, 2, 3, etc.) so that these discharge points can be referenced easily in other sections of the Storm Water Pollution Prevention Plan.

Working back from the storm water outfalls identified, now determine the drainage areas for each outfall. A topographic map can help with this task if one is readily available with the suitable scale. For larger facilities (greater than 25 acres), 7.5-minute topographic maps, available from the U.S. Geological Survey (USGS), will probably have the level of detail necessary to determine site drainage patterns.

Maps may be purchased from local commercial dealers or directly from USGS information offices. (Check local yellow pages for commercial dealers.) Topographic maps may also be purchased by mail. Standard topographic quadrangles cost around $2.50. Maps can be ordered from the following locations:

	For Alaska maps:
USGS Map Sales	USGS Map Sales
Box 25286	101 12th Ave., #12
Denver, CO 80225	Fairbanks, AK 99701

For smaller sites, examination of a topographic map may not reveal very much about the drainage patterns of the site. A simple alternative is to examine the contours of the site. A visual observation of flows or the use of small floatables or dyes in concentrated flows are simple methods to determine drainage patterns at the facility. Drainage patterns may be very obvious in some cases, such as drainage down a particular hill on the site. In areas where the site appears to be relatively flat, a rough study of storm water flow during a rain event should provide a sufficient sense of the flow patterns.

Structural Storm Water Controls. Other features to include on the site map are the locations and identification of any existing structural control measures already in place that are used to control or direct storm water runoff. A structural control measure is any physically constructed feature onsite that is specifically used to change the way that storm water flows or that is used to remove pollutants from storm water. Examples of structural controls include: retention/detention ponds, flow diversion structures (including ditches and culverts), vegetative swales, porous pavement, sediment traps, and any soil stabilization or erosion control practices. See Chapters 8 and 9 for a more complete description and illustrations of these structures. Each structure should be clearly identified on the site map.

Surface Waters. On the site map, all surface water bodies should be labeled on or next to the site. This includes any stream, river, lake, or other water body. Each water body should be identified by name. If you do not know the name of the water body, you can check the USGS topographical maps discussed above for the legal name. Your municipal government may also have municipal maps that identify small streams by name. If storm water runoff flows into a small, unnamed tributary, the name of the downstream water body will be sufficient.

Potential Pollutant Sources. To develop a useful site map for the facility's Storm Water Pollution Prevention Plan, some additional items need to be identified on the map to show what activities are taking place in each drainage area, and thus, what types of pollutants may be present in storm water from these areas. These features include:

* Topography of site.

* Location of exposed significant materials.

* Locations of past spills and leaks.

* High-risk waste generating areas and activities common on industrial sites, such as:

    - Fueling stations
    - Vehicle and equipment maintenance
    - Vehicle and equipment washing
    - Loading and unloading areas
    - Above-ground liquid storage tanks
    - Industrial waste management areas and outside manufacturing
    - Outside storage of raw materials, by-products, or finished products.

Specific BMPs may be applied to control the amount of pollutants in storm water discharges from these areas (see Chapter 7).

5.3.3 Types of Pollutants and Flow Direction

As part of the description of potential pollutant sources, the general permit regulations require a prediction of the direction of flow and an identification of the types of pollutants which are likely to be present in storm water discharges associated with industrial activity for each area of the facility that generates storm water discharges associated with industrial activity with a reasonable potential for containing significant amounts of pollutants. Factors to consider include the toxicity of chemicals; quantity of chemicals used, produced or discharged; the likelihood of contact with storm water; and history of significant leaks or spills of toxic or hazardous pollutants. Flows with a significant potential for causing erosion must be identified.[80]

5.3.4 Inventory of Exposed Materials

The next step in the Pollutant Source Assessment Phase is to conduct a material inventory at the site, specifically to look for materials that have been exposed to storm water and measures that have been taken to prevent the contact of these materials with storm water. Maintaining an up-to-date material inventory is an efficient way to identify which materials are handled at the site and which materials are contributing to storm water contamination problems. These potential pollutant sources should be identified on the facility's site map.

The pollution prevention plan provisions of EPA's general permit require that the permittee inventory materials that may be exposed to storm water at the site. The permittee must include a narrative description of the following information:[81]

* Significant materials that have been handled, treated, stored, or disposed in a manner to allow exposure to storm water between the time of three years prior to the date of permit issuance and the present.

---

[80] Id., Part IV.D.2.a.(2), 57 Fed. Reg. 44438, 44447 (Sept. 25, 1992).
[81] Id., Part IV.D.2.b., 57 Fed. Reg. 44438, 44447 (Sept. 25, 1992).

* Method(s) and location of on-site storage or disposal.

* Materials management practices employed to minimize contact of these materials with storm water runoff between the time of three years prior to the date of the issuance of the permit and the present.

* Existing structural and nonstructural control measures to reduce pollutants in storm water runoff, including their locations.

* Any treatment of storm water runoff.

Inventory of Exposed "Significant Materials". "Significant materials," as defined in 40 C.F.R. 122.26(hi(12), are substances related to industrial activities such as process chemicals, raw materials, fuels, pesticides, and fertilizers. When these substances are exposed to storm water runoff, they may be carried to a receiving stream with the storm water flow. Therefore, identification of these materials helps to determine where a potential for contamination exists and is the first step in identifying appropriate BMPs to address this contamination potential.

When compiling the inventory of exposed materials, focus on areas where the facility stores, processes, transports, or transfers any materials that are used or produced during industrial processes. Check any storage tanks, pipes, or pumping areas and note any leaks or spills. Observe any loading or unloading operations and indicate whether any industrial materials are exposed to storm water during those processes. Look at any unsealed dumpsters or disposal units/areas where the facility deposits wastes from industrial activities and document instances where waste materials are exposed to rain. Also pay attention to material-handling equipment, including everything from vehicles to pallets, where raw and waste materials from industrial activities are exposed. Finally, consider areas where particles are emitted from air vents and are likely to fall within site drainage areas, such as the roof.

Existing Management Measures and Treatment of Storm Water Runoff. After describing the potential pollutant sources in storm water runoff from the site, describe management practices currently in use. Management practices can be as simple as scheduled sweeping of the material transfer area. In this section of the plan, describe both structural and nonstructural management practices. Structural management practices are those practices that entail construction of manmade structures such as berms, detention ponds, or grassed swales, whereas nonstructural management practices involve regularly scheduled actions (such as sweeping, inspections, or improved materials handling and management practices).

Finally, describe any treatment being provided for the storm water discharges from the site. The treatment of storm water is often accomplished through holding in a detention pond which allows for settling of inorganic solids and partial removal of organic contaminants. In the case of detention ponds, describe the size and average depth of each pond on the site (storage volume). Also provide any design criteria (i.e., design flow rates, etc.) for the pond that may be available from engineering design reports or diagrams. If some of the storm water is directed into a process water treatment system, identify the type of treatment provided, and whether it is allowed under your NPDES or other discharge permit. In any case, be sure to specify areas from which the treated storm water drains.

5.3.5 Site Checklist

You may wish to use the following site checklist to help in identifying exposed materials.

[] Does your facility show signs of poor housekeeping (cluttered walkways, unswept floors, uncovered materials, etc.)?

[] Are there spots, pools, puddles, or other traces of oil, grease, or other chemicals on the ground?

[] Is there discoloration, residue, or corrosion on the roof or around vents or pipes that ventilate or drain the work area?

[] Do you see leaking equipment pipes, containers, or lines?

[] Are there areas where absorbent materials (cat litter, saw dust, etc.) are regularly used?

[] Do you notice signs such as smoke, dirt, or fumes that indicate material losses?

[] Do you smell strange odors, or experience eye, nose, or throat irritation when you first enter the work area? (These are indications of equipment leaks.)

[] Do storage containers show signs of corrosion or leaks?

[] Are there open containers, stacked drums, shelving too small to properly handle invents, or other indications of poor storage procedures?

[] Are containers properly labeled?

These are some basic guidelines meant to help you determine what kinds of things to look out for. This list does not necessarily cover every possible source of pollutants. As the site operator, you are responsible for knowing the particular concerns associated with your activity. Be as detailed as you can in your description of the significant materials exposed at your facility; discuss what you found in your assessment, the amounts present, and their location. Update this inventory whenever new, significant materials are introduced and exposed on site so that your management practices can be modified to suit any changes.

Next, you should give closer scrutiny to areas where you store or dispose of industrial materials. Inspect your various containers carefully and note whether there are any openings, holes or leaks that allow storm water to contact significant materials in those containers.

5.3.6 Spills and Leaks

The pollution prevention plan provisions of EPA's general permit require that the permittee include a list of significant spills and significant leaks of toxic or hazardous pollutants that occurred at areas that are exposed to precipitation or that otherwise drain to a storm water conveyance at the facility after the date of three years prior to the effective date of the permit. Such list must be updated as appropriate during the term of the permit.[82]

EPA has defined "significant spills" to include releases within a 24-hour period of hazardous substances in excess of reportable quantities under CWA Section 311 and CERCLA Section 102. Reportable quantities are set amounts of hazardous substances in pounds, gallons, or other units and are listed in 40 C.F.R. Part 117 and 40 C.F.R. Part 302. If the facility releases hazardous substances to the environment in reportable quantities, the owner or operator of the facility must immediately notify the National Response Center at (800) 424-8802.[83]

5.3.7 Nonstorm Water Discharges

The pollution prevention plan provisions of EPA's general permit require that the plan include a certification that all storm water outfalls have been tested or evaluated for the presence of nonstorm water discharges. The certification must include:

* Identification of potential nonstorm water discharges.

* A description of the results of any test and/or evaluation for the presence of nonstorm water discharges.

* The evaluation criteria or test method used.

* The date of testing and/or evaluation.

* The on-site drainage points that were directly observed during the test and/or evaluation.

---

[82] Id., Part IV.D.2.c., 57 Fed. Reg. 44438, 44447 (Sept. 25, 1992).

[83] For detailed information on reporting of spills and releases, see Dennison, Environmental Reporting, Recordkeeping, and Inspections: A Compliance Guide for Business and Industry (Van Nostrand Reinhold, 1995).

This certification must be signed in accordance with Part VII.G.1. of the general permit and must be included in the Storm Water Pollution Prevention Plan.[84]

If this certification is not feasible because the facility does not have access to an outfall, manhole, or other point of access to the final storm water discharge point(s), describe why the certification was not feasible. For facilities that begin industrial activities after October 1, 1992, the facility also must notify the permitting authority, within 180 days after submitting the Notice of Intent (NOI), of any potential sources of nonstorm water discharges to the storm water discharge and explain why the facility could not perform the test for nonstorm water discharges. This certification must be signed in accordance with Part VII.G.1. of the general permit and submitted to the permitting authority.

Generally, except for flows from fire fighting activities, all sources of nonstorm water that are authorized by your storm water discharge permit must be identified in the Storm Water Pollution Prevention Plan. Where necessary to minimize pollutants in these discharges, pollution prevention measures should be adopted and implemented.[85]

Examples of nonstorm water discharges include any water used directly in the manufacturing process (process water), air conditioner condensate, noncontact cooling water, vehicle wash water, or sanitary wastes. Connections of nonstorm water discharges to a storm water collection system are common, yet are often unidentified. Those types of discharges are significant sources of water quality problems.

Authorized Nonstorm Water Discharges. Unless permitted by a NPDES permit, nonstorm water discharges are illegal. If such connections are discovered, disconnect them or submit a NPDES permit application (Form 2C for process wastewater or Form 2E for nonprocess wastewater) to the permitting authority. EPA's

---

[84] See Final NPDES General Permits for Storm Water Discharges Associated With Industrial Activity, Part IV.D.3.g.(1), 57 Fed. Reg. 44438 (Sept. 25, 1992).

[85] Id., Part IV.D.3.g.(2), 57 Fed. Reg. 44438 (Sept. 25, 1992).

general permit does, however, authorize the following types of nonstorm water discharges:

* Discharges from fire-fighting activities.

* Fire hydrant flushings.

* Potable water sources, including waterline flushings.

* Irrigation drainage.

* Lawn watering.

* Uncontaminated groundwater.

* Foundation or footing drains where flows are not contaminated with process materials.

* Discharges from springs.

* Routine exterior building wash down which does not use detergents or other compounds.

* Pavement wash waters where spills or leaks of toxic or hazardous materials have not occurred and where detergents are not used.

* Air conditioning condensate.

Be sure to examine your facility's storm water permit to determine whether it authorizes any of these or other nonstorm water discharges. If your permit does not authorize nonstorm water discharges occurring at your facility, you should contact your permitting authority or the Storm Water Hotline at (703) 821-4823 for more information about how to address these discharges.

To check for nonstorm water discharges, one of several common dry weather tests may be used: (1) visual inspection; (2) plant schematic review; and (3) dye testing.

Visual Inspection. The easiest method for detecting nonstorm water connections into the storm water collection system is simply to observe all discharge points during dry weather. Inspect each discharge point on three separate occasions. As a rule, the discharge point should be dry during a period of extended dry weather, since a storm water collection system should only collect storm water. Keep in mind, however, that drainage of a particular rain event can continue for three days or more after the rain has stopped. In addition, infiltration of ground water into the underground collection system is also common. To be sure about the source of any flow during dry weather, you may need to perform one of the additional tests described below.

Plant Schematic. A review of a plant schematic is another simple way to determine if there are any interconnections into the on-site storm water collection system. A sewer map or plant schematic is a map of pipes and drainage systems used to carry process wastewater, noncontact cooling water, air conditioner condensate, and sanitary wastes (bathrooms, sinks, etc.). A common problem, however, is that sites often do not have accurate, up-to-date schematics. If you do have an accurate and reliable plant schematic, you can simply examine the pathways of the different water circuits listed above. Also be sure to investigate where the floor drains discharge. These are commonly connected to the storm sewer system, especially in older buildings.

Dye Testing. Another method for detecting improper connections to the storm water collection system is dye testing. A dye test can be performed by simply releasing a dye into either your sanitary or process wastewater system and examining the discharge points from the storm water collection system for discoloration.[86]

[86] A detailed description of the equipment needed and proper procedures for a dye test is provided in Section 4.2.5.

5.3.8 Sampling Data

Part IV.D.2.d. of EPA's general permit requires that the permittee provide a summary of existing discharge sampling data describing pollutants in storm water discharges from the facility, including a summary of sampling data collected during the term of the permit.[87]

Storm water sampling data provide information that describes the quality of storm water discharges. These data are valuable because they indicate the potential environmental risk of the discharge by identifying the types and amounts of pollutants present. In addition, these data can be used to identify potential sources of storm water pollution.

During the site assessment phase, permittees should collect and summarize any storm water sampling data that were collected in the past. Historical storm water monitoring data may be very useful in locating areas which have previously contributed pollutants to storm water discharges and identifying what the problem pollutants are. When summarizing these data, describe the sample collection procedures used. Be sure to cross reference the particular storm water outfall sampled to one of the outfalls designated on the site map.

Although some permittees may not have to conduct storm water sampling under the permit that is issued to that facility, incorporation of these data into the Storm Water Pollution Prevention Plan as it is collected will provide a basis for evaluating the effectiveness of the plan. Under EPA's general permit, certain classes of facilities are required to conduct storm water sampling either annually or semiannually throughout the term of the permit.[88] State-issued storm water general permits may include similar provisions. Generally, where sampling is required, facilities must collect and analyze grab and composite samples in accordance with the protocol established in 40 C.F.R. Part 136.[89]

---

[87] See Final NPDES General Permits for Storm Water Discharges Associated With Industrial Activity, Part IV.D.2.d., 57 Fed. Reg. 44438, 44447 (Sept. 25, 1992).

[88] See Section 4.1.4.

[89] See Chapter 4 for a discussion of the technical requirements of storm water sampling.

5.3.9 Risk Identification and Summary of Pollutant Sources

Part IV.D.2.e. of EPA's general permit requires that the permittee provide a narrative description of potential pollutant sources from the following activities: loading and unloading operations; outdoor storage activities; outdoor manufacturing or processing activities; significant dust or particulate generating processes; and on-site waste disposal practices. The description must specifically list any significant potential source of pollutants at the site and for each potential source, any pollutant or pollutant parameter (e.g., biochemical oxygen demand, etc.) of concern must be identified.[90]

After completion of each of the steps in the pollutant source assessment outlined in the preceding subsections, there should be enough information to determine which areas, activities, or materials may contribute pollutants to storm water runoff from the facility site. With this information, the facility can select the most appropriate BMPs to prevent or control pollutants from these areas.

The following is an example of how to analyze the information gathered and decide how to correct the problems identified:

> In a particular drainage area, you have a vehicle maintenance facility area where oil filters are stored outdoors. You found that no material management practices were currently being used to protect the used filters from contact with storm water. You would then suspect that the storm water draining from that area would most likely contain a significant amount of oil and grease. Therefore, you have concluded that you need to do something to reduce the possibility of oil and grease mixing with storm water.

This is the type of narrative description required by EPA's general permit. It summarizes any potential source of storm water pollutants and identifies the types of pollutants that have

[90] See Final NPDES General Permits for Storm Water Discharges Associated With Industrial Activity, Part IV.D.2.e., 57 Fed. Reg. 44438, 44447 (Sept. 25, 1992).

already been or may be found in storm water runoff from the site.

5.4 BEST MANAGEMENT PRACTICES IDENTIFICATION

After having identified and assessed potential and existing sources of contamination to storm water at the facility, the next step is to select the proper measures or Best Management Practices (BMPs) that will eliminate or reduce pollutant loadings in storm water discharges from the facility site.

As part of the CWA strategy to eliminate discharges of pollutants to receiving waters, NPDES permit limitations have become more stringent. Industrial and municipal facilities subject to the NPDES program may need to undertake additional measures to meet these permit limitations, as well as the goals of the CWA. Such measures can be technologically and economically achievable through the development of formalized plans that contain BMPs and pollution prevention practices.

5.4.1 Best Management Practices and Pollution Prevention

BMPs are inherently pollution prevention practices. Traditionally, BMPs have focused on good housekeeping measures and good management techniques intending to avoid contact between pollutants and water media as a result of leaks, spills, and improper waste disposal. However, based on the authority granted under the regulations, BMPs may include the universe of pollution prevention encompassing production modifications, operational changes, materials substitution, materials and water conservation, and other such measures.

EPA endorses pollution prevention as one of the best means of pollution control. In 1990, the Pollution Prevention Act was enacted and set forth a national policy that "pollution should be prevented or reduced at the source whenever feasible; pollution that cannot be prevented should be recycled in an environmentally safe manner, whenever feasible; pollution that cannot be prevented or recycled should be treated in an environmentally safe manner whenever feasible; and disposal or other release into the environment should be employed only as a last resort and should be conducted in an environmentally safe manner." The intent of pollution prevention practices and BMPs are similar and they can be concurrently developed in a technologically sound and cost-effective manner.

Many facilities currently implement successful measures to reduce and control environmental releases of all types of pollutants. These measures have been successfully implemented both formally as part of BMP plans and informally as part of unwritten standard operating procedures. In the context of the NPDES permit program, permittees are required to develop BMP plans to address specific areas of concern. The BMP plan developed by the permittee becomes an enforceable condition of the permit.

BMPs are developed as part of the NPDES permitting requirements to control releases of harmful substances. BMPs may apply to an entire site or be appropriate for discrete areas of an industrial facility. Many of the same environmental controls promoted as part of a BMP plan may currently be used by industry in Storm Water Pollution Prevention Plans, SPCC plans, OSHA safety programs, fire protection programs, insurance policy requirements, or standard operating procedures. Additionally, where facilities have developed pollution prevention programs, controls such as source reduction and recycling/reuse may be similar to those promoted as part of a BMP plan.

Traditionally, BMP activities have focused on activities associated with or ancillary to industrial manufacturing or treatment processes. These have been identified in CWA Section 304(e) as "plant site runoff, spillage or leaks, sludge or waste disposal, and drainage from raw material storage which the EPA Administrator determines are associated with or ancillary to the industrial manufacturing or treatment process." These activities have historically been found to be amenable to control by BMPs. Some examples include the following:

* *Material storage areas* for toxic, hazardous, and other chemicals including raw materials, intermediates, final products, or by-products. Storage areas may be piles of materials or containerized substances. Typical storage containers could include liquid storage vessels ranging in size from large tanks to 55-gallon drums; dry storage in bags, bins, silos, and boxes; and gas storage in tanks and vessels.

The storage areas can be open to the environment, partially enclosed, or fully contained.

* *Loading and unloading operations* involving the transfer of materials to and from trucks or rail cars, including in-plant transfers. These operations include pumping of liquids or gases from truck or rail car to a storage facility or vice versa, pneumatic transfer of dry chemicals during vehicle loading or unloading, transfer by mechanical conveyor systems, and transfer of bags, boxes, drums, or other containers from vehicles by forklift, hand, or other materials handling methods.

* *Facility runoff* generated principally from rainfall on a plant site. Runoff can become contaminated with harmful substances when it comes in contact with material storage areas, loading and unloading areas, in-plant transfers areas, and sludge and other waste storage/disposal sites. Fallout, resulting from plant air emissions that settle on the plant site, may also contribute to contaminated runoff. In addition to BMPs, facility runoff from industrial sites may also be directly regulated under the NPDES storm water permitting program.

* *Sludge and waste storage and disposal areas* including landfills, pits, ponds, lagoons, and deep-well injection sites. Depending on the construction and operation of these sites, there may be a potential for leaching of toxic pollutants or hazardous substances to groundwater, which can eventually reach surface waters. In addition, liquids may overflow to surface waters from these disposal operations.

With the increasing awareness of pollution prevention opportunities as well as the increase in legislation and regulatory policies directing efforts towards pollution prevention, much of the traditional focus of BMP activities is being redirected from ancillary activities to industrial manufacturing processes. This redirection has resulted in the integrated application of traditional BMPs and pollution prevention practices into cohesive and encompassing plans that cover all aspects of industrial facilities.

5.4.2 Types of BMPs

BMPs may be divided into general BMPs, applicable to a wide range of industrial operations, and facility-specific (or process-specific) BMPs, tailored to the requirements of an individual site. General BMPs are widely practiced measures that are independent of chemical compound, source of pollutant, or industrial category. General BMPs are also referred to as baseline practices, and are typically low in cost and easily implemented. General BMPs are practiced to some extent at almost all facilities. Common general BMPs include good housekeeping, preventive maintenance, inspections, security, employee training, and recordkeeping and reporting.

Facility-specific BMPs are measures used to control releases associated with individually identified toxic and hazardous substances and/or one or more particular ancillary sources. Facilityspecific BMPs are often developed when a facility notes a history of problem releases of toxic or hazardous chemicals, or when facility personnel believe that actual or potential pollutant discharge problems should be addressed. Facility-specific BMPs may include many different practices such as source reduction and on-site recycle/reuse.

Facility-specific BMPs will vary from site to site depending upon site characteristics, industrial processes, and pollutants. For example, a site-specific BMP in the form of area dikes may be adopted due to the location of the facility: facilities in flat areas or on slopes are likely to utilize dikes to control spills, whereas there may be no need for dikes for facilities located in basins. Additionally, plants handling and storing large amounts of liquid chemicals would be more likely to utilize dikes than facilities storing and using dry chemicals. Facilities experiencing erosion and sediment control problems may consider establishing vegetative buffer strips or indigenous ground cover for purposes of soil stabilization and infiltration of runoff. Facility sites located adjacent to other industrial areas may consider runon controls to prevent extraneous spills and contaminated runon from entering the facility site. Other site-specific considerations, such as endangered species, may motivate facilities to store materials in an alternate location so as to prevent exposure.

Processes also drive the determination of appropriate specific BMPs. Materials-handling procedures that expose employees to toxic chemicals (e.g., hand drawing) may prompt the consideration of procedures that reduce the potential for exposure (e.g., automated pneumatic pumping in enclosed conduits). Some examples of process-specific BMPs include the following:

* Using splash plates designed to prevent spills at a metal-finishing facility.

* Installing solvent recovery equipment to control benzene releases at a petroleum refinery.

* Purchasing solvents in reusable containers rather than 55-gallon drums to ease storage concerns and reduce wastewater resulting from requirements to triple-rinse drums.

Pollutant characteristics such as volatility and toxicity also affect BMP selection. More and more, harmful chemicals are being considered for replacement with less toxic alternatives, or for elimination from the process. Ozone layer-depleting solvents which are used for cleaning at many facilities are being replaced with detergent-based cleaning agents. Additionally, facilities using materials with toxic properties have been inspired to take more proactive control measures (e.g., double-walled containment).

5.5 BASELINE BMPs

General or "baseline" BMPs are relatively simple to evaluate and adopt. As previously indicated, general BMPs are practiced to some extent at all facilities. The pollution prevention plan provisions of EPA's general permit require that the permittee develop a description of storm water management controls appropriate for the facility, and implement such controls. The appropriateness and priorities of controls in a plan must reflect identified potential sources of pollutants at the facility. The description of storm water management controls must address the following minimum components, including a schedule for implementing such controls:[91]

* *Good housekeeping:* A program by which the facility is kept in a clean and orderly fashion.

* *Preventive maintenance:* A program focused on preventing releases caused by equipment problems, rather than repair of equipment after problems occur.

* *Visual inspections:* A program established to oversee facility operations and identify actual or potential problems.

* *Spill prevention and response:* A program designed to avoid releases due to accidental or intentional entry.

* *Sediment and erosion control:* A program designed to identify structural, vegetative, and/or stabilization measures to control significant soil erosion caused by topography, site activities, or other factors.

* *Management of runoff:* A program designed to evaluate the appropriateness of traditional storm water management practices used to divert, infiltrate, reuse, or otherwise manage storm water runoff in a manner that reduces pollutants in storm water discharges from the site.

* *Employee training:* A program developed to instill in employees an understanding of the BMP plan.

* *Recordkeeping and reporting:* A program designed to maintain relevant information and foster communication.

The purpose of this section is to describe each of these baseline BMPs and offer guidelines for selecting more

---

[91] Id., Part IV.D.3.a.-i., 57 Fed. Reg. 44438, 44447-44448 (Sept. 25, 1992).

"advanced" BMPs that are tailored to the specific pollutant sources at a particular site.

5.5.1 Good Housekeeping

As part of the pollution prevention plan requirements of EPA's general permit, the facility must implement good housekeeping measures. Good housekeeping requires the maintenance of areas which may contribute pollutants to storm water discharges in a clean, orderly manner.[92]

Good housekeeping is essentially the maintenance of a clean, orderly work environment. Maintaining an orderly facility means that materials and equipment are neat and well kept to prevent releases to the environment. Maintaining a clean facility involves the expeditious remediation of releases to the environment. Together, these terms, clean and orderly, define good housekeeping.

Maintaining good housekeeping is the heart of a facility's overall pollution control effort. Good housekeeping cultivates a positive employee attitude and contributes to the appearance of sound management principles at a facility. Some of the benefits that may result from a good housekeeping program include ease in locating materials and equipment; improved employee morale; improved manufacturing and production efficiency; lessened raw, intermediate, and final product losses due to spills, waste, or releases; fewer health and safety problems arising from poor materials and equipment management; environmental benefits resulting from reduced releases of pollution; and overall cost savings.

Good housekeeping measures can be easily and simply implemented. Some examples of commonly implemented good housekeeping measures include the orderly storage of bags, drums, and piles of chemicals; prompt cleanup of spilled liquids to prevent significant runoff to receiving waters; expeditious sweeping, vacuuming, or other cleanup of accumulations of dry chemicals to prevent them from reaching receiving waters; and proper disposal of toxic and hazardous wastes to prevent contact with and contamination of storm water runoff.

---

[92] Id., Part IV.D.3.a., 57 Fed. Reg. 44438, 44447 (Sept. 25, 1992).

The primary impediment to a good housekeeping program is a lack of thorough organization. To overcome this obstacle, a three-step process can be used, as follows:

1. Determine and designate an appropriate storage area for every material and every piece of equipment.

2. Establish procedures requiring that materials and equipment be placed in or returned to their designated areas.

3. Establish a schedule to check areas to detect releases and ensure that any releases are being mitigated.

The first two steps act to prevent releases that would be caused by poor housekeeping. The third step acts to detect releases that have occurred as a result of poor housekeeping.

As with any new or modified program, the initial stages will be the largest hurdle; ultimately, though, good housekeeping should result in savings that far outweigh the efforts associated with initiation and implementation. Generally, a good housekeeping plan should be developed in a manner that creates employee enthusiasm and thus ensures its continuing implementation.

The first step in creating a good housekeeping plan is to evaluate the facility site organization. In most cases, a thorough release identification and assessment has already generated the needed inventory of materials and equipment and has determined their current storage, handling, and use locations. This information, together with that from further assessments, can then be used to determine if the existing locations of materials and equipment are adequate in terms of space and arrangement.

Cramped spaces and those with poorly placed materials increase the potential for accidental releases due to constricted and awkward movement in these areas. A determination should be made as to whether materials can be stored in a more organized and safer manner (e.g., stacked, stored in bulk as opposed to individual containers, etc.). The proximity of materials to their place of use should also be evaluated. Equipment and materials used in a particular area should be stored nearby for convenience, but should not hinder the movement of workers or equipment. This is especially important for waste products. Where waste conveyance is not automatic (e.g., through chutes or pipes), waste receptacles should be located as close as possible to the waste generation areas, thereby preventing inappropriate disposal leading to environmental releases.

Appropriately designated areas (e.g., equipment corridors, worker passageways, dry chemical storage areas) should be established throughout the facility. The effective use of labeling is an integral part of this step. Signs and adhesive labels are the primary methods used to assign areas. Many facilities have developed innovative labeling approaches, such as color coding the equipment and materials used in each particular process. Other facilities have stenciled outlines to assist in the proper positioning of equipment and materials.

Once a facility site has been organized in this manner, the next step is to ensure that employees maintain this organization. This can be accomplished through explaining organizational procedures to employees during training sessions, distributing written instructions, and most importantly, demonstrating by example.

Support of the program must be demonstrated, particularly by responsible facility personnel. Shift supervisors and others in positions of authority should act quickly to initiate activities to rectify poor housekeeping. Generally, employees will note this dedication to the good housekeeping program and will typically begin to initiate good housekeeping activities without prompting. Although initial implementation of good housekeeping procedures may be challenging, these instructions will soon be followed by employees as standard operating procedures.

Despite good housekeeping measures, the potential for environmental releases remains. Thus, the final step in developing a good housekeeping program involves the prompt identification and mitigation of actual or potential releases. Where potential releases are noted, measures designed to prevent release can be implemented. Where actual releases are occurring, mitigation measures such as those described below may be required.

It may not always be possible to immediately correct poor housekeeping. However, deviations should occur only in emergencies. The routines and procedures established as a part of the program should allow for adequate time to conduct good housekeeping activities.

Mitigation practices. Mitigation practices are simple in theory: the immediate cleanup of an environmental release lessens chances of spreading contamination and lessens impacts due to contamination. When considering choices for mitigation methods, a facility must consider the physical state of the material released and the media to which the release occurs. Generally, the ease of implementing mitigation actions should also be considered. For example, crushed stone, asphalt, concrete, or other covering may top a particular area. Consideration as to which substance would be easier to clean in the event of a release should be evaluated.

Conducting periodic inspections is an excellent method to verify the implementation of good housekeeping measures. Inspections and related concerns are discussed in Section 5.5.3.

Good Housekeeping Measures:

* Integrate a recycling/reuse and conservation program in conjunction with good housekeeping. Include recycle/reuse opportunities for common industry wastes such as paper, plastic, glass, aluminum, and motor oil, as well as facility-specific substances such as chemicals, used oil, dilapidated equipment, etc., into the good housekeeping program. Provide reminders of the need for conservation measures including turning off lights and equipment when not in use, moderating heating/cooling, and conserving water.

* When reorganizing, keep pathways and walkways clear with no protruding containers.

* Create environmental awareness by celebrating Earth Day (April 22) and/or developing a regular (e.g., monthly) good housekeeping day.

* Develop slogans and posters for publicity. Involve employees and their families by inviting suggestions for slogans and allowing children to develop the facility's good housekeeping posters.

* Provide suggestion boxes for good housekeeping measures.

* Develop a competitive program that may include company-wide competition or facilitywide competition. Implement an incentive program to spark employee interest (i.e., $\frac{1}{2}$ day off for the shift which best follows the good housekeeping program).

* Conduct inspections to determine the implementation of good housekeeping. These may need to be conducted more frequently in areas of most concern.

* Pursue an ongoing information exchange throughout the facility, the company, and other companies to identify beneficial good housekeeping measures.

* Maintain necessary cleanup supplies (i.e., gloves, mops, brooms, etc.).

* Set job performance standards which include aspects of good housekeeping.

* Do not allow rubbish or other waste to accumulate. Properly dispose of waste or arrange to have it removed in a timely fashion.

* Do not limit good housekeeping measures to industrial locations. Office areas should also be involved in the good housekeeping program.

5.5.2 Preventive Maintenance

As part of the pollution prevention plan requirements of EPA's general permit, the facility must implement preventive maintenance (PM) measures. A PM program must involve timely inspection and maintenance of storm water management devices (e.g., cleaning oil/water separators, catch basins) as well as inspecting and testing facility equipment and systems to uncover conditions that could cause breakdowns or failures resulting in discharges of pollutants to surface waters, and ensuring appropriate maintenance of such equipment and systems.[93]

PM is a method of periodically inspecting, maintaining, and testing plant equipment and systems to uncover conditions which could cause breakdowns or failures. As part of a BMP plan, PM focuses on preventing environmental releases. Most facilities have existing PM programs. It is not the intent of the Storm Water Pollution Prevention Plan to require development of a redundant PM program. Instead, the objective is to expand the current PM program to include storm water considerations, especially the upkeep and maintenance of storm water management devices. The pollution prevention team should evaluate the existing plant PM program and recommend any necessary changes.

A PM program accomplishes its goals by shifting the emphasis from a repair maintenance system to a PM system. It should be noted that in some cases, existing PM programs are limited to machinery and other moving equipment. The PM program prescribed to meet the goals of the Storm Water Pollution Prevention Plan includes all other items (man made and natural) used to contain and prevent releases of toxic and hazardous materials. Ultimately, the well-operated PM program devised to support the pollution prevention plan should produce environmental benefits of decreased releases to the environment, as well as reduce total maintenance costs and increase the efficiency and longevity of equipment, systems, and structures.

In terms of pollution prevention plans, the PM program should prevent breakdowns and failures of equipment, containers, systems, structures, or other devices used to handle the toxic or hazardous chemicals or wastes. To meet this goal, a PM program should include a suitable system for evaluating equipment, systems, and structures; recording results; and

[93] Id., Part IV.D.3.b., 57 Fed. Reg. 44438, 44447 (Sept. 25, 1992).

facilitating corrective actions. A PM program should, at a minimum, include the following activities:

* Identification of equipment, systems, and structures to which the PM program should apply.

* Determination of appropriate PM activities and the schedule for such maintenance.

* Performance of PM activities in accordance with the established schedule.

* Maintenance of complete PM records on the applicable equipment, systems, and structures.

Generally, the PM program is designed to prevent and/or anticipate problems resulting from equipment and structural failures. However, it is unrealistic to expect that the PM program will avert the need for repair maintenance as a result of unanticipated problems. Adjustments and repair of equipment will still be necessary where problems occur, and replacement of equipment will be necessary when adjustment and repair are not sufficient.

Generally, all good PM programs will consist of the four components noted above. However, it is of particular importance that the PM program address those areas and pollutants identified during the pollutant source assessment.

Although creating and implementing PM programs sounds easy, it is often impeded by lack of funding and of organization. Lack of funding must be overcome by a facility's commitment to its PM program based on the simple truth that PM is less costly than replacement. Lack of organization can be overcome by better planning, which can be achieved by following the steps to developing an effective PM plan discussed below.

At the outset of a PM program, an inventory should be devised. This inventory should provide a central record of all equipment and structures, including: location; identifying information such as serial numbers and facility equipment numbers/names; size, type, and model; age; electrical and mechanical data; the condition of the equipment/structure; and the manufacturer's address, phone number, and person to contact. In addition to the equipment inventory, an inventory of the structures and other nonmoving parts to which the PM program is to apply should also be determined.

Inventories can be developed through inspections and/or reviews of facility specifications and operations and maintenance manuals. In some cases, it is effective to label equipment and structure with assigned numbers/names and some of the identifying information. This information may be useful to maintenance personnel in the event of an emergency situation or unscheduled maintenance where maintenance information is not readily available. Several different methods are effective for recording inventory information, including the use of index cards, prepared forms and checklists, or a computer database.

Since the PM program involves the use of maintenance materials (i.e., spare parts, lubricants, etc.), some additional considerations may apply. First, good housekeeping measures, as discussed in Section 5.5.1, are particularly important for organizing maintenance materials and keeping areas clean. A tracking system may also be necessary for organizing maintenance materials. The inventory should include information such as materials/parts description, number, item specifications, ordering information, vendor addresses and phone numbers, storage locations, order quantities, order schedules, and costs. A large facility may require a parts catalog to coordinate such information. Large facilities may also find it necessary to develop a purchase order system which maintains the stock in adequate number and in the proper order by keeping track of the minimum and maximum number of items required to make timely repairs, parts that are vulnerable to breakage, and parts that have a long delivery time or are difficult to obtain.

Once the inventory is completed, the facility should determine the PM requirements including schedules and specifications for lubrication, parts replacement, equipment and structural testing, maintenance of spare parts, and general observations. The selected PM activities should be based on the facility-specific conditions, but should be at least as stringent as the manufacturer's recommendations. Manufacturer's specifications can generally be found in brochures and pamphlets accompanying equipment. An operations and maintenance manual also may contain this information. If these sources are not available, the suggested manufacturer's recommendation can be obtained directly from the manufacturer. In cases of structures or nonmoving parts, the facility will need to determine appropriate maintenance activities (e.g., integrity testing). As with inventory information, PM information should be recorded in an easily accessible format.

After establishment of the materials inventory and the development of PM requirements, a facility should schedule and carry out PM on a regular basis. Personnel with expertise in maintenance should be available to conduct maintenance activities. In a small facility where one person may conduct regular maintenance activities, specialized contractors may supplement the maintenance program for more complex activities. An up-to-date list of outside firms available for contract work beyond the capability of the facility staff should be readily available. Additionally, procedures explaining how to obtain such support should be provided in the pollution prevention plan. Larger facilities should have sufficient PM expertise within the staff, including a PM manager, an electrical supervisor, a mechanical supervisor, electricians, technicians, specialists, and clerks to order and acquire parts and maintain records. Ongoing training and continuing education programs may be used to establish expertise in deficient areas.

Maintenance activities should be coordinated with normal plant operations so that any shutdowns do not interfere with production schedules or environmental protection. The maintenance supervisory staff should also consider other timing constraints such as the availability of the PM staff for both regularly scheduled PM and unanticipated corrective repairs.

The final step in the development of a PM program involves the organization and maintenance of complete records. A PM tracking system which includes detailed upkeep, cost, and staffing information should be utilized. A PM tracking system assists facilities in identifying potential equipment or structural problems resulting from defects, general old age, inappropriate maintenance, or poor engineering design; preparation of a maintenance department budget; and deciding whether a piece of equipment or a structure should continue to be repaired or replaced.

There are many commercial software systems that enable facilities to track maintenance. Computer systems allow for input of inventory and PM information and generate daily, weekly, monthly, and/or yearly maintenance sheets which include the required item to be maintained, the maintenance duties, and materials to be used (e.g., oil, spare parts, etc.). The system can be continually updated to add information gathered during maintenance activities. Some of the maintenance information that proves useful includes the work hours spent, materials used, frequency of downtime for repairs, and costs involved with maintenance activities. This information in turn can generate budgets and determinations of the cost effectiveness of repair versus replacement, etc. Computerized systems for maintenance tracking are usually most effective at larger facilities.

Useful manual systems may involve index cards, maintenance logs, and a maintenance schedule. Initially, inventory and PM information can be recorded on index cards. This information can be consulted during maintenance activities. Maintenance logs should also be developed for each piece of equipment and each structure, and should contain information such as the maintenance specifications and data associated with the completion of maintenance activities. Maintenance personnel should complete relevant information including the date maintenance was conducted, hours spent on duties, materials used, worker identification, and the nature of the problem.

Preventive Maintenance Measures:

* When attaching information to equipment and structures, use bold and bright colors consistent with the approaches described for good housekeeping.

* Consider and discuss additional PM procedures beyond those normally recommended by the manufacturer.

* Conduct extensive safety training for PM personnel.

* Coordinate scheduling of PM activities with facility or unit downtime.

* Keep track of how long materials have been stored. This will support an evaluation of the integrity of storage containers.

* Develop a PM staff team approach including team names (i.e., the A-team) to create enthusiasm.

* Utilize blackboards and charts to assist in organizing and conveying an annual PM schedule.

* Do not forget to stock important replacement parts and any specialized tools required to repair equipment.

* Do not create a paperwork nightmare. Develop the minimum number of well-organized logs necessary to maintain information.

* Do not let untrained, unskilled personnel conduct PM activities. Employees taking part in the PM program must be familiar with equipment and maintenance procedures.

* Do not forget to determine the availability and time needed to obtain vital parts or contractor assistance.

EPA's general permit contains additional preventive maintenance inspection requirements for facilities subject to EPCRA Section 313 reporting for water priority chemicals.[94] For these facilities, all areas of the facility must be inspected for the following at appropriate intervals as specified in the plan:

* Leaks or conditions that would lead to discharges of Section 313 water priority chemicals.

[94] Id., Part IV.D.7.b.(7)., 57 Fed. Reg. 44438 (Sept. 25, 1992).

* Conditions that could lead to direct contact of storm water with raw materials, intermediate materials, waste materials, or products.

* Examine piping, pumps, storage tanks and bins, pressure vessels, process and material-handling equipment, and material bulk storage areas for leaks, wind blowing, corrosion, support or foundation failure, or other deterioration or noncontainment.

These inspections must occur at intervals based on facility design and operational experience, and the timing must be specified in the plan. When a leak or other threatening condition is found, corrective action must be taken immediately or the facility unit or process must be shut down until the problem is repaired.

5.5.3 Visual Inspections

As part of the pollution prevention plan requirements of EPA's general permit, the facility must perform visual inspections. In addition to or as part of the comprehensive site evaluation required under Part IV.4 of the permit, qualified facility personnel must be identified to inspect designated equipment and areas of the facility at appropriate intervals specified in the plan. A set of tracking or follow-up procedures must be used to ensure that appropriate actions are taken in response to the inspections. Records of inspections must also be maintained.[95]

Inspections provide an ongoing method to detect and identify sources of actual or potential environmental releases. Inspections also act as oversight mechanisms to ensure that selected BMPs are being implemented. Inspections are particularly effective in evaluating the good housekeeping and PM programs previously discussed.

Security scans, site reviews, and facility walk-throughs conducted by plant managers and other such personnel qualify as inspections. These types of reviews, however, are often limited in scope and detail. To ensure that the objectives of the

[95] Id., Part IV.D.3.d., 57 Fed. Reg. 44438, 44447 (Sept. 25, 1992).

pollution prevention plan are being met, these types of reviews should be conducted concurrently with periodic, in-depth inspections as part of a comprehensive inspection program.

Inspections implemented as part of the pollution prevention plan should cover those equipment and facility areas identified during the pollutant source assessment as having the highest potential for environmental releases. Since inspections may vary in scope and detail, an inspection program should be developed to prevent redundancy while still ensuring adequate oversight and evaluation.

A pollution prevention inspection program should set out guidelines for the following:

1. The scope of each inspection;

2. The personnel assigned to conduct each inspection;

3. The inspection frequency;

4. The format for reporting inspection findings; and

5. Remedial actions to be taken as a result of inspection findings.

Despite the different requirements of each type of inspection, the focus of inspections conducted as part of the pollution prevention plan should not vary.

An inspection program's goal will be to ensure thoroughness while preventing redundancy. Ultimately, this will ensure that the use of resources is optimized. In addition, it should be clear that the inspection team's efforts are directed to support the operating groups in carrying out their responsibilities for equipment and personnel safety and work quality, and to ensure that all standards are met. In achieving these goals, written procedures discussing the scope, frequency and scheduling, personnel, format, and remediation procedures should be provided.

The scope of each inspection type should be discussed in the written procedures. Many different types of inspections are conducted as part of the inspection program. Guidelines for the scope of these inspections include:

* *Security scan:* Search for leaks and spills which may be occurring. Specifically examine problem areas which have been identified by the plant manager or equivalent persons. Security scans can be conducted various times daily.

* *Walk-through:* Conduct oversight of the duties associated with a security scan. In addition, ensure that equipment and materials are located in their appropriate positions. Walk-through inspections can be conducted once per shift to once per week.

* *Site review:* Conduct oversight of duties associated with a walk-through. Additionally, evaluate the effectiveness of the PM, good housekeeping, and spill prevention and response programs by visual oversight of their implementation. Site reviews can be conducted once per week to once per six months.

* *Pollution prevention plan oversight inspection:* Conduct oversight of duties associated with a site review. Evaluate the implementation of all aspects of the written pollution prevention plan, including the review of the records generated as part of these programs (e.g., inspection reports, PM activity logs). Plan oversight inspections can be conducted once per month to once per year.

* *Pollution prevention plan evaluation/reevaluation inspection:* Conduct an evaluation/reevaluation of the facility and determine the most appropriate BMPs to control environmental releases that may contaminate storm water discharges. BMP plan reevaluation inspections can be conducted once per year to once every five years.

An appropriate mix of these types of inspections should be developed based on facilityspecific considerations. The proper frequency for conducting inspections will vary based on the type of the inspection and other facility-specific factors.

Every facility is different, so it is up to the facility owner/operator to determine which areas of the facility could potentially contribute pollutants to storm water runoff, and to devise and implement an inspection program based on this information. The following list identifies some types of equipment and plant areas to include in the facility's inspection plan:

* Areas around equipment, including around pipes, pumps, storage tanks and bins, pressure vessels, pressure release valves, process and material-handling equipment, and storm water management devices.

* Areas where spills and leaks have occurred in the past.

* Material storage areas (tank farms, drum storage).

* Outdoor material-processing areas.

* Material-handling areas (loading, unloading, transfer).

* Waste generation, storage, treatment, and disposal areas.

Individuals qualified to assess the potential for environmental releases should be assigned to conduct formal inspections. Members of the pollution prevention team can generally fulfill this requirement, but they may not be available to conduct all inspections. Thus, it may be appropriate to identify and train personnel to conduct specific types of inspections. For example, shift foremen and other equivalent supervisory personnel may appropriately conduct walk-throughs and site reviews as a result of their position of authority and ability to require prompt correction if problems are observed. Personnel with immediate responsibility for an area should not be asked to conduct inspections of that area, as they may be tempted to overlook problems.

Different perspectives are useful when conducting inspections. By developing a team inspection approach or by alternating inspectors, facilities can receive a more thorough review. One inspector may observe something that another will overlook, and an inspector tends to focus on the areas with which he/she is most familiar.

An inspection checklist of areas to inspect with space for a narrative report is a helpful tool when conducting inspections. A standard form helps ensure inspection consistency and comprehensiveness. Checklists may not be necessary for each inspection performed. This may be particularly true for facilities conducting frequent inspections (once per hour, once per shift, etc.); procedures for using inspection checklists should be reasonable to prevent generation of excessive paperwork.

The findings of inspections will be useless unless they are brought to the attention of appropriate personnel and subsequently acted upon. To ensure that reports are acted upon in an expeditious and appropriate manner, procedures for routing and review of reports should be developed and followed.

Despite the usefulness of written reports, in no way should a written report replace verbal communication. Where a problem is noted, particularly environmental releases currently occurring or about to occur, it should be verbally communicated by the inspector to the responsible personnel as soon as possible.

Inspection records should note when inspections were done, who conducted the inspection, what areas were inspected, what problems were found, and steps taken to correct any problems, including who has been notified. Many industrial facilities will already have some sort of incident reporting procedure in place. Existing incident reporting and security surveillance procedures could easily be incorporated into the Storm Water Pollution Prevention Plan. These records should be kept with the plan. EPA's general permit requires that records be kept until at least one year after coverage under the permit expires.

Inspection Measures:

* Encourage workers to conduct visual inspections and report any actual or potential problems to the appropriate personnel.

* Develop inspection checklists for each type of inspection. Vary them where necessary for each part of the facility subject to BMPs.

* Consider utilizing nonregulatory support from EPA, States, or university-supported resources when conducting site assessments.

* Do not rely solely on the use of a checklist for inspections. Narrative descriptions should be included in the reports to ensure that problems are identified and discussed.

* Do not conduct inspections and then fail to provide feedback of findings of concern to the person responsible for the area inspected.

5.5.4 Spill Prevention and Response

As part of the pollution prevention plan requirements of EPA's general permit, the facility must implement spill prevention and response procedures. Areas where potential spills which can contribute pollutants to storm water discharges can occur, and their accompanying drainage points must be identified clearly in the Storm Water Pollution Prevention Plan. Where appropriate, specifying material-handling procedures, storage requirements, and use of equipment such as diversion valves in the plan should be considered. Procedures for cleaning up spills must be identified in the plan and made available to the appropriate personnel. The necessary equipment to implement a cleanup should be available to personnel.[96]

Spills and leaks together are one of the largest industrial sources of storm water pollutants and, in most cases, are avoidable. Establishing standard operating procedures, such as safety and spill prevention procedures, along with proper employee training, can reduce these accidental releases. Avoiding spills and leaks is preferable to cleaning them up after they occur, not only from an environmental standpoint, but also because spills cause increased operating costs and lower productivity.

Development of spill prevention and response procedures is a very important element of an effective Storm Water Pollution

[96] Id., Part IV.D.3.c., 57 Fed. Reg. 44438, 44447 (Sept. 25, 1992).

Prevention Plan. A spill prevention and response plan may have already been developed in response to other environmental regulatory requirements. If the facility already has a spill prevention and response plan, it should be evaluated and revised if necessary to address the objectives of the Storm Water Pollution Prevention Plan.

As part of the Pollutant Source Assessment Phase of developing the Storm Water Pollution Prevention Plan, the facility should have created a list or inventory of materials that are handled, used, and disposed of. A site map indicating the drainage area of each storm water outfall should also be created. Now overlay the drainage area map with the locations of these areas and activities with high material spill potential to determine where spills are likely to occur. Spill potential also depends on how materials are handled, the types and volumes of materials handled, and how materials are stored at the site. These factors must be described in the plan.

The activities and areas where spills are likely to occur at the site are listed and described below.

* Loading and unloading areas.

* Storage areas.

* Process activities.

* Dust- or particulate-generating processes.

* Waste disposal activities.

Loading and unloading areas have a high spill potential because the nature of the activity involves transfer of materials from one container to another. The spill potential is affected by the integrity of the container, the form of the chemical being transferred, the design of the transfer area, the proximity of this area to the storage area, and procedures for loading and unloading. Evaluate the spill potential from all loading and unloading equipment, such as barges, railroad cars, tank trucks, and front-end loaders, as well as storage and vehicle washing areas.

Storage areas, both indoor and outdoor, are potential spill areas. Outdoor storage areas are exposed to storm water runoff and may provide direct contact between potential pollutants and storm water. Indoor storage areas may contaminate storm water if the drains in the storage area are connected to the storm sewer or if improper cleanup procedures are used in the event of a spill. This evaluation should consider the type, age, and condition of the storage containers and structures, including tanks, drums, bags, and bottles. An evaluation of the spill potential of storage areas should also focus on how employees handle materials.

All process areas are potential sources of storm water contamination if the floor drains in these areas are connected to storm sewers. If these drains cannot be sealed, the process area should be evaluated for the adequacy of spill control structures, such as secondary containment, if necessary. Normal housekeeping procedures should also be considered. Some process areas are hosed down periodically and the resulting wash water contains pollutants. Outdoor process activities may contaminate storm water if spills are diverted to the storm sewer.

In addition, evaluate the spill potential from the following stationary facilities:

* Manufacturing areas.
* Warehouses.
* Chemical processing and/or blending areas.
* Temporary and permanent storage sites.
* Power-generating facilities.
* Food-processing areas.
* Tank farms.
* Service stations.

* Parking lots.

* Access roads.

Also evaluate the possibility of storm water contamination from underground sources, such as tanks and pipes. Leaking underground storage tanks are often a source of storm water contamination. Finally, in addition to identifying these and other potential spill areas, projecting the possible spill volume and type of material is critical to developing the correct response procedures for a particular area.

Through the process of developing various spill scenarios, ideas will emerge for eliminating or minimizing the spill or its impact. These solutions should be prioritized and adopted according to conditions of effectiveness, cost, feasibility, and ease of implementation. The following is a list of some suggested activities and alterations that may be made to reduce the potential that spills will occur or impact storm water quality:

* Develop ways to recycle, reclaim, and/or reuse process materials to reduce the volume brought into the facility.

* Install leak detection devices, overflow controls, and diversion berms.

* Disconnect drains from processing areas that lead to storm sewer (however, be sure that any such action would not create a health hazard within the facility).

* Adopt effective housekeeping practices.

* Adopt a materials flow/plant layout plan (i.e., do not store bags that are easily punctured near high-traffic areas where they may be hit by moving equipment or personnel).

* Perform regular visual inspections to identify signs of wear on tanks, drums, containers, storage shelves, and berms and to identify sloppy housekeeping or other clues that could lead to potential spills.

* Perform preventive maintenance on storage tanks, valves, pumps, pipes, and other equipment.

* Use filling procedures for tanks and other equipment that minimize spills.

* Use material transfer procedures that reduce the chance of leaks or spills.

* Substitute less toxic or nontoxic materials for toxic materials.

* Ensure appropriate security.

In the event that spill prevention measures fail, a swiftly executed response may prevent contamination of storm water. Spill response plans are required for various reasons by numerous programs. However, this may be the first time that a spill response plan specifically addresses protection of storm water quality.

Past experience has shown that the single most important obstacle to an effective spill response plan is its implementation. Develop a plan with its ease of implementation in mind. The spill response procedures should be clear, concise, step-by-step instructions for responding to the spill events at a particular facility. Organize the plan to facilitate rapid identification of the appropriate set of procedures. For example, the plan for a particular facility may work best when organized by spill location. Another possible method of organization is by spilled material. The key component to implementation is the ability of employees to use the plan quickly and effectively. The specific approach taken by a given facility will depend on the specific facility conditions, such as size, the number of employees, and the spill potential at the site.

Development of the spill response plan is based on the spill potential scenarios identified. It reflects a consideration of the potential magnitude and frequency of spills, of the types of materials spilled, and the variety of potential spill locations. Specific procedures may be needed to correspond to particular chemicals used, processed, stored, or otherwise handled at the site. At all times during operation of a facility, personnel should be available who have appropriate training and authority to respond to spills.

The response plan should describe the following:

* Identification of the spill response "team" responsible for implementing the spill response plan.

* Safety measures.

* Procedures to notify appropriate authorities providing assistance, such as police, fire, hospital, POTW, and so on.

* Spill containment, diversion, isolation, cleanup.

* Spill response equipment, including:

  - Safety equipment, such as respirators, eye guards, protective clothing, fire extinguishers, and two-way radios.
  - Cleanup equipment, including brooms, barriers, sweeps, adsorbents, and containers.

Following any spills, evaluate how the prevention plan was successful or unsuccessful in responding and how it can be improved.

Finally, EPA's general permit sets forth more specific requirements for facilities that are subject to EPCRA Section 313 reporting for water priority chemicals.[97] Whenever a leak or spill of a Section 313 water priority chemical occurs, the contaminated soil, material, or debris must be removed promptly and disposed of in accordance with federal, state, and local requirements and as described in the Storm Water Pollution Prevention Plan. These facilities are also required to designate a person responsible for spill prevention, response, and reporting procedures.

[97] Id., Part IV.D.7.b.(7)., 57 Fed. Reg. 44438 (Sept. 25, 1992).

5.5.5 Sediment and Erosion Control

As part of the pollution prevention plan requirements of EPA's general permit, the facility must identify areas which, due to topography, activities, or other factors, have a high potential for significant soil erosion, and identify structural, vegetative, and/or stabilization measures to be used to limit erosion.[98]

There may be areas at the facility site that are prone to soil erosion due to construction activities, steep slopes, sandy soils, or other reasons. Construction activities typically remove grass and other protective ground covers, resulting in exposure of the underlying soil to wind and rain. Similarly, steep slopes or sandy soils may not be able to hold plant life so that these soils become exposed. Because the soil surface is unprotected, dirt and sand particles are easily picked up by wind and/or washed away by rain. This process is known as erosion. Erosion can be controlled or prevented with the use of certain BMPs. A number of these measures are described in Chapter 9.

5.5.6 Management of Runoff

As part of the pollution prevention plan requirements of EPA's general permit, the facility's plan must contain a narrative consideration of the appropriateness of traditional storm water management practices (practices other than those which control the generation or source(s) of pollutants) used to divert, infiltrate, reuse, or otherwise manage storm water runoff in a manner that reduces pollutants in storm water discharges from the site. The plan must provide that measures that the permittee determines to be reasonable and appropriate must be implemented and maintained. The potential of various sources at the facility to contribute pollutants to storm water discharges associated with industrial activity must be considered when determining reasonable and appropriate measures.[99]

Appropriate measures may include vegetative swales and practices, reuse of collected storm water (such as for a process or as an irrigation source), inlet controls (such as oil/water separators), snow management activities, infiltration devices, and

[98] Id., Part IV.D.3.h., 57 Fed. Reg. 44438, 44447-44448 (Sept. 25, 1992).
[99] Id., Part IV.D.3.i., 57 Fed. Reg. 44438, 44447-44448 (Sept. 25, 1992).

wet detention/retention devices. These and other storm water management practices are described in Chapters 6 to 9.

5.5.7 Employee Training

As part of the pollution prevention plan requirements of EPA's general permit, the facility must perform employee training. Employee training programs must inform personnel responsible for implementing activities identified in the Storm Water Pollution Prevention Plan or otherwise responsible for storm water management at all levels of responsibility of the components and goals of the Storm Water Pollution Prevention Plan. Training should address topics such as spill response, good housekeeping, and material management practices. The pollution prevention plan must identify periodic dates for such training.[100]

Employee training is essential to effective implementation of the Storm Water Pollution Prevention Plan. The purpose of a training program is to teach personnel at all levels of responsibility the components and goals of the Storm Water Pollution Prevention Plan. When properly trained, personnel are more capable of preventing spills, responding safely and effectively to an accident when one occurs, and recognizing situations that could lead to storm water contamination.

Spill prevention and response procedures are described in detail in Section 5.5.4. Discuss these procedures or plans in the training program in order to ensure all plant employees, not just those on the spill response teams, are aware of what to do if a spill occurs. Specifically, all employees involved in the industrial activities at the facility should be trained about the following measures:

* Identify potential spill areas and drainage routes, including information on past spills and causes.

* Reporting spills to appropriate individuals, without penalty (e.g., employees should be provided "amnesty" when they report such instances).

---

[100] Id., Part IV.D.3.e., 57 Fed. Reg. 44438, 44447-44448 (Sept. 25, 1992).

* Specifying material-handling procedures and storage requirements.

* Implementing spill response procedures.

On-site contractors and temporary personnel should also be informed of the plant operations and design features in order to help prevent accidental discharges or spills from occurring.

Also, teach facility personnel how to maintain a clean and orderly work environment. Section 5.5.1 outlines the steps for practicing good housekeeping. Emphasize these points in the good housekeeping portion of the training program:

* Require regular vacuuming and/or sweeping.

* Promptly clean up spilled materials to prevent polluted runoff.

* Identify places where brooms, vacuums, sorbents, foams, neutralizing agents, and other good housekeeping and spill response equipment are located.

* Display signs reminding employees of the importance and procedures of good housekeeping.

* Discuss updated procedures and report on the progress of practicing good housekeeping at every meeting.

* Provide instruction on securing drums and containers and frequently checking for leaks and spills.

* Outline a regular schedule for housekeeping activities to allow you to determine that the job is being done.

* Neatly organize materials for storage.

* Identify all toxic and hazardous substances stored, handled, and produced on site. Discuss handling procedures for these materials.

Here are some suggestions of training tools to include in the facility's training program:

* Employee handbooks.
* Films and slide presentations.
* Drills.
* Routine employee meetings.
* Bulletin boards.
* Suggestion boxes.
* Newsletters.
* Environmental excellence awards or other employee incentive programs.

Providing employees with incentives, such as awards for practicing pollution prevention, is a good way to motivate personnel in working to achieve the goals of the Storm Water Pollution Prevention Plan.

Examine your plan to determine how often you should train the employees at your facility. Frequency should take into account the complexity of your management practices and the nature of your staff, including staff turnover and changes in job assignments. Facilities are required to specify a schedule for periodic training activities in their plan. In any case, regularly evaluate the effectiveness of training efforts. In many cases, this will simply involve speaking with employees to verify that information has been communicated effectively.

EPA's general permit contains additional training requirements for employees and contractor personnel that work in areas where EPCRA Section 313 water priority chemicals are used or stored.[101] These individuals must be trained in the following areas at least once per year:

* Preventive measures, including spill prevention and response and preventive maintenance.

* Pollution control laws and regulations.

* The facility's Storm Water Pollution Prevention Plan.

* Features and operations of the facility which are designed to minimize discharges of Section 313 water priority chemicals, particularly spill prevention procedures.

Employee Training Tips:

* Show strong commitment and periodic input from top management to the employee training program to create the necessary interest for a successful program.

* Ensure that announcements of training events are posted well in advance and include the times and dates of the sessions, the names and positions of the instructors, the lesson plans, and the subject material covered.

* Make the training sessions interesting. Use film and slide presentations. Bring in speakers to demonstrate the use of cleanup materials or equipment. Contact the State health and environmental agencies or the regional EPA office for films, volunteer speakers, or other training aids.

* Use employee incentive programs or environmental excellence awards to reinforce training programs.

---

[101] Id., Part IV.D.7.b.(7)., 57 Fed. Reg. 44438 (Sept. 25, 1992).

* Conduct demonstrative hands-on field training to show the effectiveness of good housekeeping, PM, or inspection programs.

* Give frequent refresher courses and consider pop quizzes to keep employees sharp.

* Do not provide training to permanent facility employees only. Overlooking temporary and contractor personnel can increase the possibility of environmental releases.

* Do not allow training session attendance to be optional. Employees in the positions that incur the most stress in terms of meeting schedules should be reminded to avoid taking shortcuts when handling toxic or hazardous chemicals.

* Do not become too standardized. Reusing an annual employee training session will be tedious to employees. Integrate new information and improve on old information.

5.5.8 Recordkeeping and Reporting

As part of the pollution prevention plan requirements of EPA's general permit, the facility must implement recordkeeping and reporting procedures. A description of incidents (such as spills or other discharges) along with other information describing the quality and quantity of storm water discharges must be included in the plan required under this part. Inspections and maintenance activities must be documented and records of such activities shall be incorporated into the plan.[102]

Keeping records of and reporting events that occur on site is an effective way of tracking the progress of pollution prevention efforts and waste minimization. Analyzing records of past spills, for example, can provide useful information for developing improved BMPs to prevent future spills of the same kind.

[102] Id., Part IV.D.3.f., 57 Fed. Reg. 44438, 44447-44448 (Sept. 25, 1992).

Recordkeeping and internal reporting represent good operating practices because they can increase the efficiency of the facility and effectiveness of BMPs.

A recordkeeping system set up for documenting spills, leaks, and other discharges, including discharges of hazardous substances in reportable quantities, could help the facility minimize incident recurrence, correctly respond with appropriate cleanup activities, and comply with legal requirements. The system for recordkeeping and reporting could also include any other information that would enhance the effectiveness of the Storm Water Pollution Prevention Plan. Keep track of reported incidents and follow up on results of inspections and reported spills, leaks, or other discharges.

Records should include the following, as appropriate:

* The date and time of the incident, weather conditions, duration, cause, environmental problems, response procedures, parties notified, recommended revisions of the BMP program, operating procedures, and/or equipment needed to prevent recurrence.

* Formal written reports. These are helpful in reviewing and evaluating the discharges and making revisions to improve the BMP program. Document all reports you call in to the National Response Center in the event of a reportable quantity discharge.[103]

* A list of the procedures for notifying the appropriate plant personnel and the names and telephone numbers of responsible employees. This enables more rapid reporting of and response to spills and other incidents.

Maintaining records for all inspections is an important element of any Storm Water Pollution Prevention Plan. Documenting all inspections, whether routine or detailed, is a good preventive maintenance technique, because analysis of inspection records allows for early detection of any potential

[103] For more information on reporting spills or other discharges, refer to Section 5.3.6 and 40 C.F.R. 117.3 and 40 C.F.R. 302.4.

problems. Recordkeeping also helps to devise improvements in the BMP program after inspection records have been analyzed. Recordkeeping and reporting for maintenance activities should also be a part of the plan as another preventive maintenance measure. Keeping a log of all maintenance activities, such as the cleaning of oil and grit separators or catch basins, will enable the facility to evaluate the effectiveness of the BMP program, equipment, and operation.

There are various simple techniques used to accurately document and report inspection results including the following:

* Field notebooks.

* Timed and dated photographs.

* Video tapes.

* Drawings and maps.

It is important to keep all records updated on:

* The correct name and address of the facility.

* The correct name and location of receiving waters.

* The number and location of discharge points.

* Principal products and production rates (where appropriate).

Records of spills, leaks, or other discharges, inspections, and maintenance activities must be retained for at least one year after coverage under the permit expires.

5.5.9 Advanced BMPs

In addition to the "baseline" BMPs discussed in the preceding subsections, which should routinely be incorporated into the Storm Water Pollution Prevention Plan, more "advanced" BMPs are available to eliminate or reduce storm water contamination from specific pollutant sources and activities at the site. These BMPs are discussed in detail in Chapters 6 to 9.

5.6 POLLUTION PREVENTION PLAN IMPLEMENTATION

When you started to design the facility's Storm Water Pollution Prevention Plan, you assembled the following key pieces of information:

* A list of actual and potential storm water discharge problems.

* The location of each outfall on a site map showing the drainage route from the property.

* A list of the management plans and practices that are already in place at the facility.

* Information contained in Section 5.5 on "baseline" BMPs for resolving storm water problems.

At the completion of this BMP identification phase, you should have accomplished the following:

* Reviewed your current management plans and practices to assess their effectiveness in addressing storm water discharges at the site.

* Scheduled the implementation of "baseline" BMPs and other "advanced" BMPs that were necessary to effectively eliminate storm water pollution problems at the site.

* Determined what to do about any identified, unpermitted connections of nonstorm water discharges to separate storm sewers.

At this point, you have designed your Storm Water Pollution Prevention Plan and the plan has been approved by facility management. This section will guide you through the next major phase in the Storm Water Pollution Prevention Plan process - implementation of the selected storm water BMPs.

Facilities must implement the provisions of the Storm Water Pollution Prevention Plan as a condition of EPA's general permit. The plan shall include a schedule for implementing identified storm water management controls.[104]

Implementing the facility's plan will involve several steps:

* Develop a schedule for implementation. For example, your schedule might include a deadline for putting improved housekeeping measures into practice. Should implementation involve certain types of modifications to your site (e.g., any construction), you will need to account for the time required to secure any necessary local or State permits.

* Assign specific individuals with responsibility for implementing aspects of the plan and/or monitoring implementation.

* Ensure that management approves of your implementation schedule and strategy and schedule regular times for reporting progress to management.

5.7 PLAN EVALUATION AND REVISION

Now that your Storm Water Pollution Prevention Plan has been put into action, you must keep it up to date by regularly evaluating the information you collected in the Assessment Phase and the controls you selected in the BMP Identification Phase. You will need to conduct regular site evaluations and revise the plan as needed.

5.7.1 Site Compliance Evaluations

Qualified personnel must conduct site compliance evaluations at appropriate intervals specified in the plan, but at least once a

[104]See Final NPDES General Permits for Storm Water Discharges Associated With Industrial Activity, Part IV.D., 57 Fed. Reg. 44438 (Sept. 25, 1992).

year (at least once in three years for inactive mining sites). As part of the compliance evaluation, the facility is required to:[105]

* Inspect storm water drainage areas for evidence of pollutants entering the drainage system.

* Evaluate the effectiveness of measures to reduce pollutant loadings and whether additional measures are needed.

* Observe structural measures, sediment controls, and other storm water BMPs to ensure proper operation.

* Inspect any equipment needed to implement the plan, such as spill response equipment.

* Revise the plan as needed within two weeks of inspection (potential pollutant source description and description of measures and controls).

* Implement any necessary changes in a timely manner, but at least within 12 weeks of the inspection.

* Prepare a report summarizing inspection results and follow up actions; the date of inspection and personnel who conducted the inspection; identify any incidents of noncompliance or certify that the facility is in compliance with the plan.

* All incidents of noncompliance must be documented in the inspection report. Where there are no incidents of noncompliance, the inspection report must contain a certification that the facility is in compliance with the plan.

* Sign the report in accordance with Part VII.G of the general permit and keep it with the plan.[106]

---

[105] Id., Part IV.D., 57 Fed. Reg. 44438, 44448 (Sept. 25, 1992).
[106] See Section 5.8.2.

Annual site compliance evaluations are comprehensive inspections performed by individuals specifically designated in the Storm Water Pollution Prevention Plan as having responsibility for conducting such inspections. These employees should be familiar with all facility industrial operations and Storm Water Pollution Prevention Plan goals and requirements. Furthermore, inspectors should be able to make necessary management decisions or have direct access to management.

This annual evaluation provides a basis for evaluating the overall effectiveness of your Storm Water Pollution Prevention Plan. In particular, the annual site compliance evaluation will allow you to verify that the description of potential pollutant sources contained in the plan is accurate, that the plan drainage map is accurate or has been updated to reflect current conditions, and that controls identified in the plan to reduce pollutants in storm water discharges are accurately identified, in place, and working. The annual site compliance evaluation will also identify where new controls are needed so that you may implement them and incorporate them into the plan.

The scope of the annual site compliance evaluation will depend on various factors, including the scope of the Storm Water Pollution Prevention Plan and the size and nature of the activities occurring at the facility. The process for conducting the evaluation should follow these steps:

1. Review the Storm Water Pollution Prevention Plan and draw up a list of those items which are part of material handling, storage, and transfer areas covered by the plan.

2. List all equipment and containment in these areas covered in the plan.

3. Review facility operations for the past year to determine if any more areas should be included in the original plan, or if any existing areas were modified so as to require plan modification; change plan as appropriate.

4. Conduct inspection to determine if all storm water pollution prevention measures are accurately identified in the plan, and are in place and working properly.

5. Document findings.

6. Modify Storm Water Pollution Prevention Plan as appropriate.

Since each facility and Storm Water Pollution Prevention Plan is unique, the exact inspection format will vary from facility to facility. All documentation regarding conditions necessitating modification to the Storm Water Pollution Prevention Plan should be kept on file as part of the plan until one year after coverage under the permit expires.

5.7.2 Plan Revisions

The Storm Water Pollution Prevention Plan must be amended whenever there is a change in design, construction, operation, or maintenance, which may impact the potential for pollutants to be discharged or if the Storm Water Pollution Prevention Plan proves to be ineffective in controlling the discharge of pollutants. Facilities are not required to submit a notice to the permitting authority each time the pollution prevention plan is modified unless the permitting authority specifically requests changes to be made to the plan.[107]

In order for the Storm Water Pollution Prevention Plan to be effective, the facility should ensure that its plan complies with any permit conditions that apply to the facility and that all facility features and operations have been represented accurately. Should either of these conditions not be met by the plan, the facility must make the necessary changes. Either the managers of facilities or the permitting authority may recommend changes to the plan.

Development of Storm Water Pollution Prevention Plans is based on site-specific features. When there are changes in design, construction, operation, or maintenance that will have a significant effect on the potential for discharging pollutants in storm water at a facility, the Storm Water Pollution Prevention

[107] See Final NPDES General Permits for Storm Water Discharges Associated With Industrial Activity, Part IV.C., 57 Fed. Reg. 44438, 44448 (Sept. 25, 1992).

Plan should be modified to reflect the changes and new conditions. For example, if the facility begins to use a new chemical in its production operations, proper handling procedures for this chemical should be incorporated into the facility plan.

It may also be necessary to change the plan because it has proven to be ineffective in controlling storm water contamination based on the results of routine visual inspections (see Section 5.5.2) or more comprehensive site evaluations (see Section 5.7.1).

5.8 PLAN ADMINISTRATION

This section provides guidance on some of the administrative requirements related to organizing and developing the facility's Storm Water Pollution Prevention Plan. These general requirements include:

* Deadlines for plan development and implementation.

* Who must sign the plan.

* Where to keep the plan.

* How to make changes to the plan that are required by the permitting authority.

5.8.1 Schedule for Plan Development and Implementation

EPA's general permit provides the following deadlines for development and implementation of Storm Water Pollution Prevention Plans:[108]

1. For storm water discharge associated with industrial activity that is existing on or before October 1, 1992, the plan:
   a. Shall be prepared on or before April 1, 1993 (and updated as appropriate); and

[108] Id., Part IV.A., 57 Fed. Reg. 44438, 44446 (Sept. 25, 1992).

b. Shall provide for implementation and compliance with the terms of the plan on or before October 1, 1993.

2. For any facility where industrial activity commences after October 1, 1992, but on or before December 31, 1992, the plan:

a. Shall be prepared on or before the date 60 calendar days after the commencement of industrial activity (and updated as appropriate); and

b. Shall provide for compliance with the terms of the plan on or before the date 60 calendar days after the commencement of industrial activity (and updated as appropriate).

3. For any facility where industrial activity commences on or after January 1, 1993, the plan:

a. Shall be prepared on or before the date of submission of an NOI to be covered under this permit (and updated as appropriate); and

b. Shall provide for compliance with the terms of the plan on or before the date of submission of an NOI to be covered under this permit (and updated as appropriate).

4. For storm water discharges associated with industrial activity from an oil and gas exploration, production, processing, or treatment operation or transmission facility that is not required to submit a permit application on or before October 1, 1992 in accordance with 40 C.F.R. 122.26(c)(1)(iii), but after October 1, 1992 has a discharge of a reportable quantity of oil or a hazardous substance for which notification is required pursuant to either 40 C.F.R. 110.6, 40 C.F.R. 117.21 or 40 C.F.R. 302.6, the plan:

a. Shall be prepared on or before the date 60 calendar days after the first knowledge of such release (and updated as appropriate); and

b. Shall provide for compliance with the terms of the plan and this permit on or before the date 60 calendar days after the first knowledge of such release (and updated as appropriate).

5. For storm water discharges associated with industrial activity from a facility that is owned or operated by a municipality that has participated in a timely group application where either the group application is rejected or the facility is denied participation in the group application by EPA, the plan:

   a. Shall be prepared on or before the 365th day following the date on which the group is rejected or the denial is made (and updated as appropriate); and

   b. Shall provide for compliance with the terms of the plan and this permit on or before the 545th day following the date on which the group is rejected or the denial is made.

Portions of the plan addressing additional requirements for storm water discharges from facilities subject to Parts IV.D.7 (EPCRA Section 313) and IV.D.8 (salt storage) of the general permit shall provide for compliance with the terms of the requirements identified in Parts IV.D.7 and IV.D.8 as expeditiously as practicable, but not later than October 1, 1995. Facilities which are not required to report under EPCRA Section 313 prior to July 1, 1992, must provide for compliance with the terms of the requirements identified in Parts IV.D.7 and IV.D.8 as expeditiously as practicable, but not later than three years after the date on which the facility is first required to report under EPCRA Section 313. However, plans for facilities subject to the additional requirements of Part IV.D.7 and IV.D.8 must provide for compliance with the other terms and conditions of this permit in accordance with the appropriate dates provided in Part IV.1, 2, 3, or 5 of this permit.

The permitting authority may grant a written extension for plan preparation and compliance for new dischargers (after October 1, 1992) upon showing of good cause.

The deadlines to complete and comply with or implement your facility's Storm Water Pollution Prevention Plan may depend on the type of permit under which your facility is covered. Be sure to read your permit carefully so that you know what the deadlines are. Many NPDES-delegated States may issue general permits for storm water that contain deadlines similar to the deadlines in EPA's General permits.

5.8.2 Signature Requirements

Where the facility is subject to storm water permit requirements, all reports, certifications, or information either submitted to the permitting authority or to the operator of a large or medium MS4, or required to be maintained by the permittee onsite should be signed as follows:[109]

* For a *corporation*, the plan must be signed by a "responsible corporate officer." A responsible corporate officer may be any one of the following:
  - A president, secretary, treasurer, or vice-president of the corporation in charge of a principal business function, or any other person who performs similar policy or decision-making functions for the corporation; or
  - The manager of one or more manufacturing, production, or operating facilities employing more than 250 persons or having gross annual sales or expenditures exceeding $25,000,000 (in second-quarter 1980 dollars) if authority to sign documents has been assigned or delegated to the manager in accordance with corporate procedures.

* For a *partnership or sole proprietorship*, the plan must be signed by a general partner or the proprietor, respectively.

* For a *municipality, State, Federal, or other public agency*, the plan must be signed by either:
  - The principal executive officer or ranking official, which includes the chief executive officer of the agency, or
  - A senior executive officer having responsibility for the overall operations of a principal geographic unit of the agency (e.g., Regional Administrators of EPA).

[109] Id., Part VII.G.1.-2.c., 57 Fed. Reg. 44438, 44454 (Sept. 25, 1992).

Any of the persons listed above may designate a duly authorized representative to sign for them. A person is a duly authorized representative only if:

* The authorization is made in writing by a person described above and submitted to the permitting authority; and

* The authorization specifies either an individual or a position having responsibility for the overall operation of the regulated facility or activity, such as the position of manager, operator, superintendent, or position of equivalent responsibility or an individual or position having overall responsibility for environmental matters for the company.

If an authorization is no longer accurate because a different individual or position has responsibility for the overall operation of the facility, a notice of change in authorized individual or authorized position must be made in writing and submitted to the permitting authority.

5.8.3 Certification

To ensure that the facility's Storm Water Pollution Prevention Plan is completely developed and adequately implemented, the NPDES permit will generally require that an authorized facility representative sign and certify the plan. The authorized facility representative should be someone at or near the top of the facility's management chain, such as the president, vice-president, or a production manager who has been delegated the authority to sign and certify this type of document. In signing the plan, the corporate officer is attesting that the information is true. This signature provides a basis for an enforcement action to be taken against the person signing the plan and related reports.

The permittee should be aware that CWA Section 309 provides for significant penalties where information is false or the permittee violates, either knowingly or negligently, its permit requirements. In some cases, the general permit may require certification of the plan by a professional engineer.

Specific signatory requirements will be listed in the NPDES permit.

Part VII.G.2.d. of EPA's general permit requires that any person signing documents under the permit must sign the following certification:[110]

> *"I certify under penalty of law that this document and all attachments were prepared under my direction or supervision in accordance with a system designed to assure that qualified personnel properly gathered and evaluated the information submitted. Based on my inquiry of the person or persons who manage the system, or those persons directly responsible for gathering the information, the information submitted is, to the best of my knowledge and belief, true, accurate, and complete. I am aware that there are significant penalties for submitting false information, including the possibility of fine and imprisonment for known violations."*

EPA's general permit contains additional certification requirements for facilities subject to reporting under EPCRA Section 313 for water priority chemicals.[111] The plan must be reviewed and certified by a Registered Professional Engineer and recertified every three years or as soon as practicable after significant modifications are made to the facility. This certification that the plan was prepared in accordance with good engineering practices, however, does not relieve the facility owner or operator of responsibility to prepare and implement the plan.

5.8.4 Plan Location and Public Access

Although all plans are to be maintained on site, some NPDES storm water permits may require that facilities submit copies of their Storm Water Pollution Prevention Plans to the permitting authority for review. Examine your permit carefully to determine what submittal requirements apply to your facility. Even if your permit does not require that you automatically

---

[110] Id., Part VII.G.2.d., 57 Fed. Reg. 44438, 44454-44455 (Sept. 25, 1992).

[111] Part IV.D.7.b.(10).

submit your plan to your permitting authority, you must provide copies of the plan to your permitting authority or to your municipal operator upon request. Plans and associated records are available to the public by request through the permitting authority.

5.8.5 Plan Modifications Required by the Permitting Authority

Upon reviewing your plan, the permitting authority may find that it does not meet one or more of the minimum standards established by the pollution prevention plan requirements. In this case, the permitting authority will notify you of changes needed to improve the plan.

Any changes required by the permitting authority must be made within 30 days, unless otherwise provided by the notification, and the facility must submit a signed certification (in accordance with Part VII.G.1.) that the requested changes have been made.[112]

For example, where a facility has not addressed spill response procedures for a toxic chemical to the extent that the permitting authority believes is necessary, the facility will be required to revise the procedures. The permitting authority retains the authority to make this type of request at any time during the effective period of the plan.

5.9 SPECIAL REQUIREMENTS

In addition to the minimum baseline BMPs discussed in previous sections, facilities may be subject to additional "special" requirements. Not all facilities will have to include these special requirements in their Storm Water Pollution Prevention Plan. Be sure to check your permit closely for these conditions. In particular, EPA's general permit includes special requirements for:

* Facilities that discharge storm water through municipal separate storm sewer systems.

* Facilities with salt storage piles.

[112] Id., Part IV.B.3., 57 Fed. Reg. 44438 (Sept. 25, 1992).

* Facilities subject to EPCRA, Section 313 reporting requirements.

5.9.1 Discharges Through MS4s

Permittees must comply with conditions in municipal storm water management programs developed under the NPDES permit issued for that system to which the industrial facility discharges, provided that the facility was directly notified of the applicable requirements by the municipal operator. The facility must be in compliance with these conditions by the deadlines specified in the pollution prevention plan.[113]

EPA's Phase I regulations for storm water discharge permit applications require large and medium municipal separate storm sewer systems (systems serving a population of 100,000 or more) to develop storm water management programs in order to control pollutants discharged through the municipal systems.[114] These management programs will address discharges of industrial storm water through the systems to the extent that they are harmful to the water quality of receiving streams. Municipalities should be aware of the facilities with storm water discharges associated with industrial activity that discharge into their separate storm sewer system because the Phase I regulations require these facilities to notify the municipal operator. In addition, facilities covered by general permits will typically be required to submit a copy of their NOI to the municipal operator.

EPA emphasizes that it is the facility's responsibility to inform the municipality of all storm water discharges associated with industrial activity to the separate storm sewer system. Facilities with such discharges that have not yet contacted the appropriate municipal authority should do so **immediately**.

Although facility-specific Storm Water Pollution Prevention Plans for industries are designed to prevent pollutants from entering storm water discharges, the municipal operator may find it necessary to impose specific requirements on a particular

---

[113] See Section 5.8.1 for these deadlines.

[114] National Pollution Discharge Elimination System Permit Application Regulations for Storm Water Discharges, 55 Fed. Reg. 47990 (Nov. 16, 1990).

industrial facility or class of industrial facilities in some situations. One way to ensure that facilities comply with these requirements is to include a provision in the facility's NPDES storm water discharge permit that directly requires compliance. This mechanism provides a basis for enforcement action to be directed, where necessary, against the owner or operator of the facility with a storm water discharge associated with industrial activity.

5.9.2 Salt Storage Piles

Where storm water from a salt storage pile is discharged to U.S. waters, the pile must be covered or enclosed to prevent exposure to precipitation, except when salt is being added to or taken from the pile. Discharges must comply with this provision as expeditiously as practicable, but in no event later than October 1, 1995.[115]

Facilities may use salt for deicing purposes or part of their industrial processes. Since exposed salt piles will easily contaminate storm water runoff, an obvious BMP for these piles is to cover them with a tarp or other covering or enclose them in a shed or building. This requirement may not, however, be applicable to all Storm Water Pollution Prevention Plans. Where runoff from the salt pile is not discharged to U.S. waters, then this requirement would not apply since the pollutants will not reach a water body.

5.9.3 EPCRA Section 313 Facilities

Section 313 of EPCRA requires operators of manufacturing facilities that handle toxic chemicals in amounts exceeding threshold levels (listed at 40 C.F.R. 372.25) to report to the government on an annual basis. Because these types of facilities handle large amounts of toxic chemicals, EPA concluded that they have an increased potential to degrade the water quality of receiving streams. To address this risk, EPA established specific control requirements in its general permit. In particular, these requirements apply to Section 313 facilities that report for

[115] See Final NPDES General Permits for Storm Water Discharges Associated With Industrial Activity, Part IV.D.8., 57 Fed. Reg. 44438 (Sept. 25, 1992).

"water priority chemicals" that include any of over 200 chemicals that have been identified by EPA as especially toxic to water ecosystems.

Many of the requirements outlined below are specifically designed to address the water quality concerns that toxic chemicals present. Incorporation of these requirements into site-specific Storm Water Pollution Prevention Plans will prevent spills and leaks of water priority chemicals and eliminate or reduce other opportunities for exposure of toxic chemicals to storm water, thus protecting receiving streams from toxic discharges.

The following specific control requirements must be practiced in areas where Section 313 water priority chemicals are stored, handled, processed, or transferred:

* Provide containment, drainage control, and/or diversionary structures:
  - Prevent or minimize runon by installing curbing, culverting, gutters, sewers, or other controls, and/or
  - Prevent or minimize exposure by covering storage piles.

* Prevent discharges from all areas:
  - Use manually activated valves with drainage controls in all areas, and/or
  - Equip the plant with a drainage system to return spilled material to the facility.

* Prevent discharges from liquid storage areas:
  - Store liquid materials in compatible storage containers.
  - Provide secondary containment designed to hold the volume of the largest storage tank plus precipitation.

* Prevent discharges from loading/unloading areas:
  - Use drip pans and/or
  - Implement a strong spill contingency and integrity testing plan.

* Prevent discharges from handling/processing/transferring areas:
  - Use covers, guards, overhangs, door skirts.
  - Conduct visual inspections or leak tests for overhead piping.

* Introduce facility security programs to prevent spills:
  - Use fencing, lighting, traffic control, and/or secure equipment and buildings.

Additional requirements for facilities subject to EPCRA Section 313 include baseline BMPs that have been enhanced to address specific storm water concerns associated with the handling of toxic chemicals. These additional requirements are discussed in various portions of Section 5.5.

5.10 POLLUTION PREVENTION PLAN WORKSHEETS

The remainder of this chapter contains worksheets designed to assist facilities in developing and implementing their Storm Water Pollution Prevention Plans. The following worksheets are provided:

Worksheet 1	Pollution Prevention Team
Worksheet 2	Site Map
Worksheet 3	Material Inventory
Worksheet 3a	Exposed Significant Materials
Worksheet 4	List of Significant Spills and Leaks
Worksheet 5	Nonstorm Water Discharge Assessment
Worksheet 6	Nonstorm Water Discharge Failure to Certify Form
Worksheet 7	Pollutant Source Identification
Worksheet 7a	BMP Identification
Worksheet 8	Implementation Schedule
Worksheet 9	Employee Training Program/Schedule

In addition, Appendix C contains a sample pollution prevention plan for an industrial facility, which illustrates how each of these worksheets are filled out and used in conjunction with plan development and implementation.

POLLUTION PREVENTION TEAM (Section 2.1.1) MEMBER ROSTER	Worksheet #1 Completed by: ____________ Title: ____________ Date: ____________

Leader: ____________ Title: ____________

Office Phone: ____________

Responsibilities: ____________

Members:

(1) ____________ Title: ____________

Office Phone: ____________

Responsibilities: ____________

(2) ____________ Title: ____________

Office Phone: ____________

Responsibilities: ____________

(3) ____________ Title: ____________

Office Phone: ____________

Responsibilities: ____________

(4) ____________ Title: ____________

Office Phone: ____________

Responsibilities: ____________

DEVELOPING A SITE MAP (Section 2.2.1)	Worksheet #2 Completed by: ______ Title: ______ Date: ______

Instructions: Draw a map of your site including a footprint of all buildings, structures, paved areas, and parking lots. The information below describes additional elements required by EPA's General Permit (see example maps in Figures 2.3 and 2.4).

EPA's General Permit requires that you indicate the following features on your site map:

- All outfalls and storm water discharges
- Drainage areas of each storm water outfall
- Structural storm water pollution control measures, such as:
  - Flow diversion structures
  - Retention/detention ponds
  - Vegetative swales
  - Sediment traps
- Name of receiving waters (or if through a Municipal Separate Storm Sewer System)
- Locations of exposed significant materials (see Section 2.2.2)
- Locations of past spills and leaks (see Section 2.2.3)
- Locations of high-risk, waste-generating areas and activities common on industrial sites such as:
  - Fueling stations
  - Vehicle/equipment washing and maintenance areas
  - Area for unloading/loading materials
  - Above-ground tanks for liquid storage
  - Industrial waste management areas (landfills, waste piles, treatment plants, disposal areas)
  - Outside storage areas for raw materials, by-products, and finished products
  - Outside manufacturing areas
  - Other areas of concern (specify:__________________)

MATERIAL INVENTORY (Section 2.2.2)	Worksheet #3 Completed by: ____________ Title: ____________ Date: ____________

Instructions: List all materials used, stored, or produced onsite. Assess and evaluate these materials for their potential to contribute pollutants to storm water runoff. Also complete Worksheet 3A if the material has been exposed during the last three years.

Material	Purpose/Location	Quantity (units)			Quantity Exposed in Last 3 Years	Likelihood of contact with storm water. If yes, describe reason.	Past Significant Spill or Leak	
		Used	Produced	Stored			Yes	No

DESCRIPTION OF EXPOSED SIGNIFICANT MATERIAL
(Section 2.2.2)

Worksheet #3A
Completed by: ______________________
Title: ______________________
Date: ______________________

Instructions: Based on your material inventory, describe the significant materials that were exposed to storm water during the past three years and/or are currently exposed. For the definition of "significant materials" see Appendix B of the manual.

Description of Exposed Significant Material	Period of Exposure	Quantity Exposed (units)	Location (as indicated on the site map)	Method of Storage or Disposal (e.g., pile, drum, tank)	Description of Material Management Practice (e.g., pile covered, drum sealed)

LIST OF SIGNIFICANT SPILLS AND LEAKS (Section 2.2.3)	Worksheet #4 Completed by: ______ Title: ______ Date: ______

Directions: Record below all significant spills and significant leaks of toxic or hazardous pollutants that have occurred at the facility in the three years prior to the effective date of the permit.

Definitions: Significant spills include, but are not limited to, releases of oil or hazardous substances in excess of reportable quantities.

1st Year Prior

Date (month/day/year)	Spill	Leak	Location (as indicated on site map)	Description				Response Procedure		Preventive Measures Taken
				Type of Material	Quantity	Source, If Known	Reason	Amount of Material Recovered	Material No Longer Exposed to Storm Water (True/False)	

2nd Year Prior

Date (month/day/year)	Spill	Leak	Location (as indicated on site map)	Description				Response Procedure		Preventive Measures Taken
				Type of Material	Quantity	Source, If Known	Reason	Amount of Material Recovered	Material No Longer Exposed to Storm Water (True/False)	

3rd Year Prior

Date (month/day/year)	Spill	Leak	Location (as indicated on site map)	Description				Response Procedure		Preventive Measures Taken
				Type of Material	Quantity	Source, If Known	Reason	Amount of Material Recovered	Material No Longer Exposed to Storm Water (True/False)	

NON-STORM WATER DISCHARGE ASSESSMENT AND CERTIFICATION (Section 2.2.4)

Worksheet #5
Completed by: ______________________
Title: ______________________
Date: ______________________

Date of Test or Evaluation	Outfall Directly Observed During the Test (identify as indicated on the site map)	Method Used to Test or Evaluate Discharge	Describe Results from Test for the Presence of Non-Storm Water Discharge	Identify Potential Significant Sources	Name of Person Who Conducted the Test or Evaluation

CERTIFICATION

I, ______________________ (responsible corporate official), certify under penalty of law that this document and all attachments were prepared under my direction or supervision in accordance with a system designed to assure that qualified personnel properly gather and evaluate the information submitted. Based on my inquiry of the person or persons who manage the system or those persons directly responsible for gathering the information, the information submitted is, to the best of my knowledge and belief, true, accurate, and complete. I am aware that there are significant penalties for submitting false information, including the possibility of fine and imprisonment for knowing violations.

A. Name & Official Title (type or print)	B. Area Code and Telephone No.
C. Signature	D. Date Signed

NON-STORM WATER DISCHARGE ASSESSMENT AND FAILURE TO CERTIFY NOTIFICATION (Section 2.2.4)

Worksheet #6
Completed by: ______________
Title: ______________
Date: ______________

Directions: If you cannot feasibly test or evaluate an outfall due to one of the following reasons, fill in the table below with the appropriate information and sign this form to certify the accuracy of the included information.

List all outfalls not tested or evaluated, describe any potential sources of non-storm water pollution from listed outfalls, and state the reason(s) why certification is not possible. Use the key from your site map to identify each outfall.

Important Notice: A copy of this notification must be signed and submitted to the Director within 180 days of the effective date of this permit.

Identify Outfall Not Tested/Evaluated	Description of Why Certification Is Infeasible	Description of Potential Sources of Non-Storm Water Pollution

CERTIFICATION

I certify under penalty of law that this document and all attachments were prepared under my direction or supervision in accordance with a system designed to assure that qualified personnel properly gather and evaluate the information submitted. Based on my inquiry of the person or persons who manage the system or those persons directly responsible for gathering the information, the information submitted is, to the best of my knowledge and belief, true, accurate, and complete. I am aware that there are significant penalties for submitting false information, including the possibility of fine and imprisonment for knowing violations, and that such notification has been made to the Director within 180 days of ________ (date permit was issued), the effective date of this permit.

A. Name & Official Title (type or print)	B. Area Code and Telephone No.
C. Signature	D. Date Signed

POLLUTANT SOURCE IDENTIFICATION
(Section 2.2.6)

Worksheet #7
Completed by: ________________
Title: ________________
Date: ________________

Instructions: List all identified storm water pollutant sources and describe existing management practices that address those sources. In the third column, list BMP options that can be incorporated into the plan to address remaining sources of pollutants.

Storm Water Pollutant Sources	Existing Management Practices	Description of New BMP Options
1.		
2.		
3.		
4.		
5.		
6.		
7.		
8.		
9.		
10.		

BMP IDENTIFICATION (Section 2.3.1)	Worksheet #7a Completed by: ________ Title: ________ Date: ________

Instructions: Describe the Best Management Practices that you have selected to include in your plan. For each of the baseline BMPs, describe actions that will be incorporated into facility operations. Also describe any additional BMPs [activity-specific (Chapter 3) and site-specific BMPs (Chapter 4)] that you have selected. Attach additional sheets if necessary.

BMPs	Brief Description of Activities
Good Housekeeping	
Preventive Maintenance	
Inspections	
Spill Prevention Response	
Sediment and Erosion Control	
Management of Runoff	
Additional BMPs (Activity-specific and Site-specific)	

IMPLEMENTATION
(Section 2.4.1)

Worksheet #8
Completed by: ____________
Title: ____________
Date: ____________

Instructions: Develop a schedule for implementing each BMP. Provide a brief description of each BMP, the steps necessary to implement the BMP (i.e., any construction or design), the schedule for completing those steps (list dates) and the person(s) responsible for implementation.

BMPs	Description of Action(s) Required for Implementation	Scheduled Completion Date(s) for Req'd. Action	Person Responsible for Action	Notes
Good Housekeeping	1.			
	2.			
	3.			
Preventive Maintenance	1.			
	2.			
	3.			
Inspections	1.			
	2.			
	3.			
Spill Prevention and Response	1.			
	2.			
	3.			
Sediment and Erosion Control	1.			
	2.			
	3.			
Management of Runoff	1.			
	2.			
	3.			
Additional BMPs (Actively-specific and site-specific)	1.			
	2.			
	3.			

EMPLOYEE TRAINING
(Section 2.4.2)

Worksheet #9
Completed by: ______________
Title: ______________
Date: ______________

Instructions: Describe the employee training program for your facility below. The program should, at a minimum, address spill prevention and response, good housekeeping, and material management practices. Provide a schedule for the training program and list the employees who attend training sessions.

Training Topics	Brief Description of Training Program/Materials (e.g., film, newsletter course)	Schedule for Training (list dates)	Attendees
Spill Prevention and Response			
Good Housekeeping			
Material Management Practices			
Other Topics			

Chapter 6

INDUSTRY-SPECIFIC BMPs

6.1 INTRODUCTION

Chapter 5 discussed the planning, development, and implementation of a best management practice (BMP) plan, including the scope of the BMP applicability and the components of an effective plan. The intent of this chapter is to illustrate through examples how plan components are implemented in various industries. Examples of BMPs are provided on an industry-specific basis.

Opportunities for pollution prevention through the implementation of BMPs are available for a multitude of industries. Since all industries cannot be addressed in this chapter, the scope has been narrowed to address industries that discharge the greatest quantities of the 17 priority pollutants targeted by EPA's Pollution Prevention Strategy of 1991 based on data contained in the Toxics Release Inventory database. This targeting method identifies opportunities where BMPs may be most effective in preventing the greatest quantity of water pollution by the pollutants of greatest concern. Although this chapter is limited to BMP identification for nine industry categories, the processes identified may be applicable for control of similar processes or pollutants by other industries.

This chapter gives the reader specific examples of effective best management and pollution prevention practices, as well as instances where facilities have successfully implemented such practices.

6.2 METAL-FINISHING INDUSTRY

The category of metal finishing includes manufacturers that take raw metal stock and subject it to various treatments to produce a product at, or closer to, its finished stage. Manufacturers classified as metal finishers perform similar operations that fall under a variety of Standard Industrial Classification (SIC) codes, including industries in major groups 34 (fabricated metal products), 35 (machinery, except electrical), 36 (electrical and electronic machinery, equipment, and supplies), 37 (transportation equipment), 38 (measuring, analyzing, and controlling instruments; photographic, medical, and optical goods; watches and clocks), and 39 (miscellaneous manufacturing industries).

6.2.1 Industry Processes/Wastes Generated

The processes used to treat raw metal stock and, correspondingly, the wastes produced are the common link among the metal-finishing category members. Some of these processes are especially amenable to BMPs; that is, implementation of BMPs is relatively easy and results in a significant reduction in the discharge of pollutants. Listed below are processes common among metal finishers and the targeted pollutants that enter wastewater streams:

Electroplating: Typical wastes produced include spent process solutions containing copper, nickel, chromium, brass, bronze, zinc, tin, lead, cadmium, iron, aluminum, and compounds formed from these metals.

Electroless plating: The most common wastes produced are spent process solutions containing copper and nickel.

Coating: Depending on the coating material that is being applied, wastes of concern include spent process solutions containing hexavalent chromium and active organic and inorganic solutions.

Etching and chemical milling: Typical solutions used in etching and milling that ultimately enter the waste stream and are of concern include chromic acid and cupric chloride.

Cleaning: Various organic and inorganic compounds enter the wastewater stream from cleaning operations.

The source of the targeted pollutants are process solutions and raw materials that enter the wastewater stream primarily through rinsing or cleaning processes. A work piece that is removed from a process or cleaning solution is typically subjected to rinsing directly afterwards and carrying excess process contaminants, referred to as dragout, into the rinse tank. The dragout concentrates pollutants in the rinse tank, which is typically discharged into the sewer system.

Another pathway by which targeted pollutants enter the wastewater stream is through the disposal of spent batch process solutions into the sewer system. Spent solutions consist of aqueous wastes and may contain accumulated solids as well. Spent solutions are typically bled at a controlled rate into the wastewater stream. Other sources of pollutants in wastewater streams include cleanup of spills and washdown of fugitive aerosols from spray operations.

6.2.2 Effective BMPs

Numerous practices have been developed to eliminate or minimize discharges of pollutants from the metal-finishing industry. Successful source reduction measures have been implemented to eliminate cyanide plating baths, as well as substitute more toxic solvents with less toxic cleaners. In many cases, cleaning with solvents has been eliminated altogether through the use of water-based cleaning supplemented with detergents, heating, and/or agitation. Other source reduction measures have been implemented to minimize the discharges of toxic materials. For example, drain boards and splash plates have been commonly installed to prevent drips and spills. Additionally, the design of immersion racks or baskets and the positioning of parts on these racks or baskets have also been optimized to prevent trapping of solvents, acids/caustics, or plating baths.

The utilization of recycle and reuse measures have also been commonly used. Many facilities have been able to minimize water use and conserve rinse waters and plating baths by measures including the following:

* Utilizing a dead rinse resulting in the concentration of plating bath pollutants. This solution may be reused directly or further purified for reuse.

* Conserving waters through countercurrent rinsing techniques.

* Utilizing electrolytic recovery, customized resins, selective membranes, and adsorbents to separate metal impurities from plating baths, acid/caustic dips, and solvent cleaning operations.

These operations and measures not only extend the useful life of solutions, but also prevent or reduce the discharge of pollutants from these operations.

6.2.3 Case Studies

Two industries highlighted in this section have implemented BMPs that resulted in substantial cost savings and pollutant reductions. For example, Emerson Electric implemented a program that resulted in savings of more than $700,000 per year and reductions in solvents, oxygen-demanding pollutants, and metals. BMPs implemented by a furniture manufacturer in the Netherlands resulted in a reduction in metals discharged from 945 to 37 kilograms per year and a decrease in water use from 330,000 to 20,000 cubic meters per year.

Emerson Electric, a manufacturer of power tools, implemented a Waste and Energy Management Program to identify opportunities for pollution prevention. An audit resulted in the following actions:

* Development of an automated electroplating system that reduced process chemical usage by 25%, process batch dumps by 20%, and wastewater treatment cost by 25%.

* Installation of a water-based electrostatic immersion painting system to replace a solventbased painting system. The water-based system resulted in a waste solvent reduction of more than 95%.

* Installation of an ultrafiltration system that recovers 65 lbs per day of waste oil and purifies 2,500 lbs per day of alkaline cleaning solution for reuse, which resulted in a reduction of 5-day biochemical oxygen demand loadings to the treatment system of 370 lbs per month. This avoided the need for installation of additional treatment.

* Installation of an alkaline and detergent and steam degreasing system, which resulted in a reduction in waste solvents by 80%.

In addition to the reduction of pollutants, Emerson realized annual costs savings of $642,000 in reduced raw material use, $2,200 in reduced water use, and $52,700 in reduced waste disposal.

A furniture manufacturer in the Netherlands reduced metals in its effluent by switching to cyanide-free baths, allowing for longer drip times, using spray rinsing, reusing water, and implementing a closed cooling system. These BMPs, complemented by the installation of treatment technology, reduced metals in the effluent from 945 to 37 kilograms per year. Water use also decreased from 330,000 to 20,000 cubic meters per year.

6.3 ORGANIC CHEMICALS, PLASTICS, AND SYNTHETIC FIBERS (OCPSF) MANUFACTURING

The Organic Chemicals, Plastics, and Synthetic Fibers (OCPSF) industry manufactures more than 25,000 different organic chemical, plastic, and synthetic fiber products. It includes both those facilities whose primary products are OCPSF and the facilities that use or produce these chemicals ancillary to their primary production. OCPSF manufacturers have two types of facilities - those with chemical synthesis as their primary function and those that recover organics, plastics, and synthetic fibers as by-products of other unrelated manufacturers.

OCPSF manufacturers include SIC code 2821 (plastic materials, synthetic resins, and nonvulcanizable elastomers), SIC code 2822 (cellulosic manmade fibers), SIC 2823 (synthetic organic fibers except cellulosic), SIC code 2824 (cyclic crudes and intermediates, dyes, and organic pigments), and SIC code 2869 (industrial organic chemicals not elsewhere classified). All OCPSF products are derived from the same raw materials (methane, ethane, propene, butane, higher aliphatic compounds, benzene, toluene, and xylene).

6.3.1 Industry Processes/Wastes Generated

As a result of the variety and complexity of the processes used and of products manufactured, there is an exceptionally wide variety of pollutants found in the wastewaters of this industry including conventional pollutants, metals, and miscellaneous organics resulting from product and by-product formation.

Contaminated wastewater generation occurs at a number of points, mainly as direct and indirect contact processes, equipment cooling, equipment cleaning, air pollution control systems, and storm water. Direct contact during manufacturing or processing is found in the use of aqueous reaction media. When water is used as a medium for OCPSF chemical processes, a high-strength process wastewater is produced. After the primary reaction has been completed and the final product has been separated from the water media, some residual product and unwanted by-products remain. Indirect contact process wastewaters include the recovery of solvents and volatile organics from the chemical reaction kettle. Vacuum jets utilize streams of water used to create a vacuum which draws off volatilized solvents and organics from the reaction kettle into solution. Later, recoverable solvents are typically separated and reused while unwanted volatile organics remain in solution in the vacuum water. This wastewater is discharged as process wastewater. Steam ejector systems are similar to vacuum jets, but steam is used instead of water. The steam is then drawn off and condensed, forming a source of process wastewater. Batch processing may require repeated and extensive equipment cleanup between batches, which is usually accomplished with water. Additionally, water scrubbers on emission control devices, and leaks and spills through the plant which contaminate storm water are two other contributors of potentially high concentration of pollutants to discharges.

6.3.2 Effective BMPs

Due to the individuality of many organic manufacturing operations, pollutant and processspecific source reduction and recycle/reuse measures may only be useful in relation to one facility. However, OCPSF manufacturing facilities are similar in that their organic chemicals comprise their raw materials and final products, and the generation and discharge of significant quantities of pollutants result during cleaning processes. Thus, much of the source reduction and recycle/reuse measures have focused on reducing pollutants lost to wastewater during cleanup.

Dedicated equipment has been purchased by many batch processors to avoid the loss of valuable materials and to prevent the generation of wastewater during cleanup activities. Additionally, many batch processors have provided for the capture of washdown waters for later recycle/reuse. Alternate cleaning methods such as manual wipedown with squeegees also have been shown to help recover organic products which would otherwise be lost during equipment cleanup.

6.3.3 Case Studies

Many OCPSF facilities have implemented programs that use a variety of techniques to minimize pollutant discharges from their plants, save money, and reduce health risks. For example, a paint manufacturer underwent a waste assessment to identify opportunities for implementing pollution prevention practices. Lab experiments that were conducted to test the feasibility of eliminating cleanup steps determined that each 10% decrease of waste volume saved $6,000 per year in disposal costs. In response to this finding, the plant rescheduled production to disperse pigments only before batch formulation, which eliminated the need for intermediate storage, reduced the need for solvents such as methyl ethyl ketone, and allowed for cleaning with a small amount of compatible solvent. A process redesign included the development of a production plan that produces paint from light to dark batches, eliminating intermediate cleanup steps which generate wastewater. Ultimately, savings were realized due to reductions in raw materials use, water use, and waste disposal.

Atlantic Industries' Nutley, New Jersey facility reduced water use from 750,000 gallons per day to 300,000 to 400,000 gallons per day and also have lessened discharges of organics and other pollutants. One measure that assisted the facility in achieving these reductions involved the simultaneous increase in process chemical concentration, lowering of reaction temperatures, and adoption of new methods for combining dye components. Ultimately, these and other improvements have reduced the amount of organics and inorganics in the wastewater by 50,000 and 250,000 gallons per year, respectively.

6.4 TEXTILES MANUFACTURING

Textile mills are manufacturing facilities that transform fiber into yarn, fabric, or other finished textile products. Those mills that fall under SIC major group 23 (apparel and other textile mill products) use dry processes that normally do not result in wastewater discharges. Some of the mills that fall under SIC major group 22 (textile mill products), however, use wet processing.

6.4.1 Industry Processes/Wastes Generated

Characteristics of the major wet manufacturing processes and pollutants in the wastewater discharges that may be targeted for BMPs are listed below:

* *Raw wool scouring is* the first treatment performed on wool, in which the wool is washed to remove the impurities peculiar to wool fibers. These impurities are present in great quantities, and include grease, sweat, dirt, feces, vegetable matter, disinfectants, and insecticides. It has been estimated that for every pound of fibers obtained, 1½ pounds of waste impurities are produced, mostly dirt, grit, and grease.

* *Scouring is* employed to remove natural and acquired impurities from fibers and fabric. The nature of the scouring operation is highly dependent on the fiber type. For example, cotton fabric is sometimes loaded into a pressure vessel containing a solution of sodium hydroxide, soap, and sodium silicate, after which it is completely rinsed to clean

the fibers and remove residual alkali. Synthetics, on the other hand, require only light scouring.

* *Carbonizing* removes burrs and other vegetable matter from loose wool or woven wool goods to prevent unequal absorption of dyes. The overall water requirements for the carbonization of wool may be substantial. For example, wool is carbonized using sulfuric acid, then rinsed to remove the acid. The wool is then neutralized using a sodium carbonate solution. A final rinse removes the alkalinity.

* *Fulling* gives woven woolen cloth a thick, compact, and substantial feel, finish, and appearance. To accomplish this, the cloth is mechanically worked in fulling machines in the presence of heat, moisture, and sometimes pressure. This allows the fibers to felt together, which causes shrinkage, increases the weight, and obscures the threads of the cloth. Fulling is performed by either alkali or acid methods. Fulling is followed by extensive washing to remove process chemicals and prevent rancidity and wool spoilage.

* *Desizing* removes the sizing compounds applied to the yarns and is usually the first wet finishing operation performed on woven fabric. It consists of soaking the fabric in a solution of mineral acid or enzymes and thoroughly washing the fabric.

* *Mercerizing* increases the tensile strength, luster, sheen, dye affinity, and abrasion resistance of cotton goods. It is accomplished by impregnating fabric with cold sodium hydroxide solution, an alkali solution that causes swelling of cotton fibers. In many mills, the sodium hydroxide is reclaimed in caustic recovery units and concentrated for reuse.

* *Bleaching is* a finishing process used to whiten cotton, wool, and synthetic fibers. In addition, bleaching dissolves sizing, natural pectins and waxes, and small particles of foreign matter. It is primarily accomplished with hydrogen

peroxide, although hypochlorite, peracetic acid, chlorine dioxide, sodium perforate, or reducing agents may be used.

* *Dyeing and printing* are the most complex of the wet processing operations in textile mills. Many mechanisms and many types of dyes are used in coloring textile fibers. Acid dyes are sodium salts, usually of sulfonic acids or carboxylic acids. Azoic dyes are insoluble pigments anchored within the fiber by padding with a soluble coupling compound, usually naphthol. In addition, common salt and surface active compounds are usually necessary to speed the reaction. Basic dyes are usually hydrochlorides of salts or organic bases and are most effective with acrylic fibers. Direct dyes resemble acid dyes in that they are sodium salts of sulfonic acids and are almost invariably azo compounds. Disperse dyes use several carriers such as acetic acid to color cellulose acetate. Mordant dyes have no natural affinity for textile fibers, but dye well when applied to cellulosic or protein fibers that have been mordanted with a metallic oxide such as chromium. Reactive dyes include many methods and chemicals such as sodium chloride, urea, sodium carbonate, sodium hydroxide, and tri-sodium phosphate. Sulfur dyes are complex organic compounds that are insoluble in water, but dissolve in a solution of sodium sulfide to which sodium carbonate has been added. Vat dyes, among the oldest natural coloring matters used, are insoluble in water, but become soluble when treated with reducing agents and used with chemicals such as sodium hydroxide. Final washing of the fabric to remove excess dyes and print paste, along with the cleanup of mixing tanks, applicator equipment, and belts, contributes wastewater associated with the dyeing and printing processes. This cleaning process often involves the use of solvents.

In the production of textile products, pollutants generally enter the waste stream during rinsing or cleaning operations. These pollutants may include acids and alkalis such as those used in scouring, carbonizing, fulling, and mercerizing processes. Solvents used for cleanup are predominant at times. Zinc may present a wastewater problem in yarn spinning and manufacturing.

6.4.2 Effective BMPs

BMPs have been successfully applied in the textile industry and range from wastewater recycling and reuse and chemical substitution to process modification and computerization of controls. American Enka, a yarn and thread mill, and United Piece Dye Works, a textile dye and finishing company, have achieved substantial cost savings, reduced pollutant levels in the wastewater effluent, and met permit effluent limits by the successful implementation of best management and pollution prevention practices. A discussion of their successes is presented below.

6.4.3 Case Studies

For its rayon yarn manufacturing process, American Enka redesigned and implemented a precipitation system to remove and recycle zinc. The redesigned precipitation system involves a twostage process where, in the second stage, zinc precipitates onto an existing slurry of zinc hydroxide crystals. Sulfuric acid is then used to convert the zinc hydroxide to zinc sulfate. Zinc sulfate is recycled back to the yarn spinning bath. This two-stage process has achieved an estimated savings of $383,000 per year, and removes zinc from the wastewater and solid waste streams.

United Piece Dye Works was able to meet its effluent limits for phosphorus by materials substitution in the production process without any capital expenditure. A detailed evaluation of the production processes, process chemistry, and the chemicals used identified the sources of phosphorus. Process modifications to reduce the use of phosphate chemicals, such as hexametaphosphate, and substitution of chemicals not containing phosphate were made. The use of phosphoric acid was eliminated. The level of phosphorus in the wastewater effluent was reduced from 7.7 to less than 1 mg/l through this pollution prevention practice of source reduction.

6.5 PULP AND PAPER PRODUCTS MANUFACTURING

Paper and allied products (SIC major group 26) includes manufacturers of pulp, paper and paperboard, and paper products. The six primary subcategories include pulp mills (SIC group number 261), which manufacture pulp from wood (hardwoods or softwoods) or from other materials such as rags, linters, waste paper, and straw; paper mills (SIC group number 262), which manufacture paper and paper products; paperboard mills (SIC group number 263), which manufacture paperboard and paperboard-related products; companies covered under SIC group number 264, which produce converted paper and paperboard products, such as envelopes, nontextile bags, die-cut paper, pressed and molded pulp goods, sanitary paper products, stationery, and tablets; manufacturers of paperboard containers and boxes (SIC group number 265), which produce folding paperboard boxes, corrugated and solid fiber boxes, sanitary food containers, fiber cans, tubes, drums, and similar products; and companies covered by SIC group number 266, which includes manufacturers of building paper and board from wood pulp and other fibrous materials.

6.5.1 Industry Processes/Wastes Generated

The production of pulp, paper, and paperboard involves four major processes: raw material preparation, pulping and recovery, bleaching, and papermaking. A discussion of each process and its associated wastes is provided below.

* *Raw material preparation* includes log washing, bark removal, and chipping and screening processes. These processes can require large volumes of water, but the use of dry bark removal techniques or the recycle of washwater or water used in wet barking operations reduces water consumption.

* *Pulping and recovery* reduces raw material into pulp suitable for further processing. Pulp production results in relatively large quantities of wastewater and wastewater pollutants. The wood entering the pulping process consists of cellulose fibers, lignin, semi-cellulose, and other compounds.

Lignin, a complex polymer that binds and strengthens wood fibers, is believed to contain dioxin precursors. Pulping processes vary from basic mechanical action, such as groundwood pulping, to complex chemical digesting sequences, such as in the alkaline (soda or kraft), sulfite, or semi-chemical processes.

Mechanical pulping does not involve use of chemicals; little or none of the wood material is dissolved. Thus, softwoods, which are easier to tear and grind, are typically used in this pulping process. The resultant pulps are generally used in manufacturing newsprint, catalogues, and toweling.

Chemical pulping removes lignin to enhance fiber flexibility, resulting in a stronger paper product, but lower fiber yields (40 to 50%). Sulfite pulp may be blended with mechanical pulps as a strengthener and is commonly used in production of viscose rayon, acetate fibers and films, plastic fillers, and cellophanes. Semi-chemical pulping is often used for newsprint, containers, and computer cards. Kraft pulping accounts for approximately 75% of the pulp produced for paper and paperboard due to the number of wood types that can be processed. Also, extracts released during the process such as turpentine, tall oil, and resin can be sold separately as commodity chemicals. Pulping process wastes include pulp rejects, cellulosic fines, white water, and chemical recovery wastes.

* *Bleaching* results in the removal of color caused by lignins and resins, or by spent cooking liquor from the pulp left by inefficient washing. Therefore, multistage bleaching processes are performed to produce light-colored or white products. Conventional bleaching involves five stages wherein chlorine is used as the dominant chemical in a series of alternating acid and alkaline bleaching and washing phases. Dioxins, furans, hexachlorobenzene, and hundreds of organochlorine by-products (acidic, phenolic, and neutral compounds) result from the bleaching process.

* *Papermaking* follows pulp preparation processes and encompasses further mixing and blending with noncellulosic materials to create the furnish for paper making. Further preparation steps may include dyeing, sizing, and starching to increase water resistance. The furnish involves a dilute water suspension of pulp, from which a layer of fiber is deposited on a fine screen. Finally, this layer is removed, pressed, dried, and, if desired, coated to form final products. Chemicals used may include titanium, zinc sulfate, lithophone pigments, waxes, starches, sodium silicate, glues, resins, rubber latex, and hydrocarbons. Coating operations typically involve use of chemicals such as polyvinyl chloride, polypropylene, saran lacquer, rubber, acrylic latex, styrene-butadiene latex, polyvinyl acetate, polyvinyl alcohol, and carboxymethyl cellulose.

6.5.2 Effective BMPs

Due to their toxicity and persistence in the environment, chlorinated organics such as dioxins and furans produced by the paper industry are often targeted for pollution prevention. Common practices have included:

* Discontinuing the use of pitch dispersants and defoamers which may contain chlorinated dioxin and chlorinated furan precursor compounds

* Maximizing delignification in the pulping process

* Maximizing brownstock pulp-washing efficiency

* Optimizing bleaching processes through process control monitoring and automation which introduces limited amounts of bleach at specific times

* Utilizing chlorine dioxide or hydrogen peroxide as alternatives to bleach, or in some cases eliminating bleaching altogether.

Although much of the focus has been on source reduction measures targeted at dioxin and dioxin precursor formation, recycle and reuse opportunities are also being utilized. More and more facilities are finding the benefits of recapturing pulp by in-process and final discharge treatment techniques. Facilities are also finding it more economical to utilize dirty water and treatment plant effluent for washing and other miscellaneous processes. In some cases, closed-loop systems can be achieved. Wood slivers and chips screened out during raw material preparation processes can be dewatered in a press and burned in a bark boiler. This process eliminates solid waste while generating power.

Source reduction and recycle methods used by the pulp and paper manufacturing industry have resulted in lessened wastewater and pollutant discharges, which in turn minimizes costs associated with treatment and water usage. Two specific examples of how industry implemented BMPs in the pulp and paper industry are highlighted below.

6.5.3 Case Studies

One paper mill in England modified the bleaching stage of pulp manufacturing to reduce water usage and reduce the coloration of wastes. This was accomplished by preceding the chlorine, caustic soda, and chlorine dioxide bleaching with oxygen bleaching, which reduced the quantities of reagents and water used in the conventional bleaching process. Rinse water from the oxygen bleaching can be used to wash cooked pulp, thus reducing the coloration of wastes. The mill reduced water usage by 50 cubic meters per ton of manufactured pulp. Use of caustic soda, chlorine, and chlorine dioxide was also reduced.

A closed-cycle, effluent-free bleached kraft mill in Canada reduced the amount of fresh water needed in the system by 50%. Technology innovations include modifications to the bleaching sequence; countercurrent washing to reduce the amount of fresh water needed; reuse of all bleach plant effluents in the pulp mill; removal of sodium chloride from the white liquor; use of an effluentfree process for generating chlorine dioxide; and installation of spill tanks and other minor changes to facilitate collection and reuse of water throughout the system. The estimated cost to install the system would be $4.5 million (in 1975 dollars) for a 725-ton-per-day mill, where annual savings were reported to be $2.2 million.

6.6 PESTICIDES FORMULATION

The formulation of pesticides and agricultural chemicals falls into SIC code 2879, and includes companies that formulate and prepare agricultural pest control chemicals or pesticides. In pesticide formulation, highly concentrated organics manufactured elsewhere are converted into pesticide products such as insecticides, herbicides, and fungicides that are ready for use by farmers and gardeners.

6.6.1 Industry Processes/Wastes Generated

There are three types of pesticide formulations: solvent based, water based, and solid based. Solvent-based formulations use a solvent or a solvent-water emulsion as the carrier solution for the active pesticide ingredient. Typical solvents are light aromatics such as xylene, chlorinated organics such as 1,1,1-trichloroethane, and mineral spirits. With water-based formulations, water serves as the carrier for the active pesticide ingredient. Both solvent- and water-based formulations are applied directly in liquid form or propelled as an aerosol. Solid-based pesticide formulations are prepared by blending solid active ingredients with inert solids such as clay or sand. Some dry formulations are prepared by absorbing liquid active ingredients into solid carrier materials. Examples of common solid-based formulations are dusts, wettable powders, granules, treated seed, and bait pellets and cubes.

Pesticide formulating facilities generate wastes during such operations as cleaning of mixing and storage equipment, housekeeping, and laboratory testing for quality assurance. Commonly generated wastewaters include those from equipment cleanup gathered as a result of raw materials left over in containers; pesticide dust and scrubber water from air pollution control equipment; offspecification products; laboratory wastes; spills; waste sands or clays; laundry wastewater; and contaminated storm water runoff.

The significant pollutant parameters in the pesticide industry include organic pollutants, suspended solids, pH, nutrients, the pesticides specific to the product manufactured, metals, phenol, and cyanide. The active ingredients in insecticides include inorganic compounds, organic compounds, chlorinated hydrocarbons, carbamates, and organophosphates. Herbicide formulations include phenoxy, metal organic compounds, triazine, urea, amide, benzoic, and other organic and inorganic compounds. Fungicides utilize organic and inorganic compounds.

6.6.2 Effective BMPs

Many of the circumstances surrounding the individuality of OCPSF manufactured products is shared by pesticide formulators. As such, much of the focus of pesticides formulation source reduction and recycle/reuse measures have been on pollutants released to water during equipment cleanup. Some of the commonly used practices have included:

* Use of dedicated equipment for batch processing to avoid losses of raw materials and products, thereby preventing the generation of wastewater during cleanup activities

* Capturing of washdown waters in tanks for later reuse

* Adoption of cleaning methods such as squeegee wipedown, which helps recover pesticide products which would otherwise be lost during equipment washdown.

Other effective source reduction measures have also been practiced in the pesticides formulation industry including the use of dry air pollution control devices and more controlled, efficient batch sequencing.

6.6.3 Case Studies

Dow Chemical reduced chlorinated hydrocarbons in the wastewater effluent by 98% and hydrocarbon emissions by 92%. They also reduced the volume of packaging wastes by implementing practices such as material substitution, equipment modification, process modification, housekeeping improvements, and periodic assessments of employee performance. Some of the measures implemented by Dow include the following:

* Packaging the pesticide Dursban in 4-ounce water-soluble packages instead of the 2-pound metal cans

previously used. This reduces packing waste volume and eliminates the disposal problem associated with the empty metal cans.

* Shipping the active pesticide ingredient in tank cars instead of 55-gallon drums. This reduces packing wastes and allows tank cars to be rinsed using a solvent present in the ingredient's formulation.

* Adding the drying agent utilizing computers and instream analyzers instead of manual feed and lab analysis. This reduces the wastewater discharged by 37%, the chlorinated hydrocarbon in wastewater by 98%, and the hydrocarbon emissions to air by 92%.

* Implementing other waste minimization measures such as process changes, recycling and reuse programs, and statistical analyses performed on data representing employee performance to pinpoint problem areas and minimize waste.

6.7 PHARMACEUTICALS MANUFACTURING

The pharmaceutical manufacturing industry encompasses the manufacture, extraction, processing, purification, and packaging of chemical materials to be used as medication for humans and animals. Industry products include natural substances extracted from plants and animals, chemically modified natural substances, synthetic organic chemicals, metal-organics, and inorganic materials.

The pharmaceutical industry's SIC codes include 2833 (medicinal chemicals and botanical products), 2834 (pharmaceutical preparations), 2841 (soaps and other detergents, except specialty cleaners), and 2844 (perfumes, cosmetics, and other toilet preparations). Pharmaceuticals may be manufactured by batch, continuous, and semicontinuous manufacturing operations, but batch production is the most common of these manufacturing techniques. Fermentation, extraction, chemical synthesis, and mixing/compounding/formulating are the processes used in these operations.

6.7.1 Industry Processes/Wastes Generated

The pollutants resulting from the manufacturing of pharmaceuticals are described below:

* *Fermentation:* Solvents such as methylene chloride, benzene, chloroform, acetone, ethyl acetate, and methanol are most often used in this process. Copper and zinc also are used in fermentation recovery processes. Chemical disinfectants such as compounds containing phenols are used for equipment sterilization.

* *Biological and natural extraction:* Most waste from this subcategory is solid waste. Detergents and disinfectants are also normally found in wastewater, as are solvents such as phenol, benzene, chloroform, 1,2-dichloroethane, acetone, and 1,4-dioxane. Ammonia is used to control pH.

* *Chemical synthesis:* Benzene and toluene are found in the majority of the process waste stream. Other solvents used include xylene, cyclohexane, pyridine, chloroform, and methylene chloride. Chemical synthesis also generates acids, bases, cyanides, metals, and other pollutants.

* *Mixing / compounding / formulating:* Various wastes are generated by these operations and include those pollutants found in the previous operations.

The waste streams generated during these various processes result from cleaning and sterilizing equipment, chemical spills, rejected products, and the processes themselves. The primary wastewater source is equipment water wash. Another source is small amounts of nonrecyclable waste dust that may be generated during mixing or tableting operations.

6.7.2 Effective BMPs

The pharmaceutical point source category is characterized by a low ratio of finished products to raw materials, especially among drugs produced by fermentation and natural extraction. Disposal and management of the large volumes of raw material waste present both a logistical and a financial burden.

Therefore, BMPs that minimize waste generation are important in reducing the release of pollutants into water, air, and soil. By implementing BMPs, many pharmaceutical companies have taken advantage of the dual benefits of reduced waste generation and more cost-efficient operations.

As with the OCPSF manufacturing and pesticide formulation industries, pharmaceutical manufacturing is characterized by significant quantities of pollutants and wastewater resulting from equipment cleanup. Many of the source reduction and recycle/reuse discussed in Sections 6.3.2 and 6.6.2 are also applicable to these facilities. Two successfully implemented BMPs are described below.

6.7.3 Case Studies

A pharmaceutical factory producing sophisticated biochemicals, bulk pharmaceutical compounds, and immunochemicals by batch production has considered waste minimization and management a high priority. This plant has enjoyed the benefits of carefully planned BMPs that have improved wastewater discharges and working conditions and have saved money. BMPs successfully established at this plant include the following:

* Wherever possible, the plant recovers and recycles used solvents. This process saves $292 per batch.

* Butyl acetate vapors are recovered through the dedication of a separate source of vacuum for drying product crystals. Savings from this recovery amount to $26 per batch.

The Merck Rahway, New Jersey, facility has implemented measures which have resulted in the recovery of 229,600 pounds of acetone per year which would normally be discharged. Outside of savings which can be attributed to lessened raw material expenditures, this reuse practice resulted in a reduction in sewer fees of $47,750 per year.

6.8 PRIMARY METALS MANUFACTURING

SIC major group 33, primary metal industries, includes facilities involved in smelting and refining of metals from ore, pig, or scrap; rolling, drawing, extruding, and alloying metals; manufacturing castings, nails, spikes, insulated wire, and cable; and production of coke. Major subcategories include blast furnaces, steel works, rolling and finishing mills (SIC group number 331); iron and steel foundries (SIC group number 332); primary and secondary smelters and refiners of nonferrous metals such as copper, lead, zinc, aluminum, tin, and nickel (SIC group numbers 333 and 334); establishments engaged in rolling, drawing, and extruding nonferrous metals (SIC group number 335); and facilities involved in nonferrous castings (SIC group number 336) and related fabricating operations.

6.8.1 Industry Processes/Wastes Generated

The main processes common to metal-forming operations and the wastes that are typically generated are discussed below:

* *Sintering:* This process agglomerates iron-bearing materials (generally fines) with iron ore, limestone, and finely divided fuel such as coke breeze. The fine particles consist of mill scale from hot rolling operations and dust generated from basic oxygen furnaces, open hearth furnaces, electric arc furnaces, and blast furnaces. These raw materials are placed on a traveling grate of a sinter machine. The surface of the raw materials is ignited by a gas and burned. As the bed burns, carbon dioxide, cyanides, sulfur compounds, chlorides, fluorides, and oil and grease are released as gas. Sinter may be cooled by air or a water spray at the discharge end of the machine, where it is then crushed, screened, and collected for feeding into blast furnaces. Wastewater results from sinter cooling operations and air-scrubbing devices which utilize water.

* *Ironmaking.* Molten iron is produced for steelmaking in blast furnaces using coke, iron ore, and limestone. Blast furnace operations use water for noncontact cooling of the furnace, stoves, and ancillary facilities and to clean and cool

the furnace top gases. Other water, such as floor drains and drip legs, contribute a lesser portion of the process wastewaters.

* *Steelmaking:* Steel is an iron alloy containing less than 1% carbon. Raw materials needed to produce steel include hot metal, pig iron, steel scrap, limestone, burned lime, dolomite fluorspar, and iron ores. In steelmaking operations, the furnace charge is melted and refined by oxidizing certain constituents, particularly carbon in the molten bath, to specified levels. Processes include the open hearth furnace, the electric hearth furnace, the electric arc furnace, and the basic oxygen furnace, all of which generate fumes, smoke, and waste gases. Wastewaters are generated when semiwet or wet gas collection systems are used to cleanse the furnace off gases. Particulates and toxic metals in the gases constitute the main source of pollutants in process wastewaters.

* *Casting operations:* This subcategory includes both ingot casting and continuous casting processes. Casting refers to the procedure of forming molten metal into a specified shape. Molten metal is distributed into an oscillating, water-cooled mold, where solidification takes place. As the metal solidifies into the mold, the cast product is typically cooled using water, which is subsequently discharged.

* *Forming operations:* Forming is achieved by passing metal through cylindrical rollers which apply pressure and reduce the thickness of the metal. Rolling reduces ingots to slabs or blooms. Secondary operations reduce slabs or blooms to billets, plates, shapes, strips, and other forms. Cooling and lubricating compounds are used to protect the rolls, prevent adhesion, and aid in maintaining the desired temperature. Hot rolling generates wastewaters laden with toxic organic compounds, suspended solids, metals, and oil and grease. Cold rolling operations, occurring at temperatures below the recrystallization point of the metal, require more lubrication. The lubricants used in cold rolling include more concentrated oil-water mixtures, mineral oil, kerosene-based lubricants (neat oils), or graphite-based lubricants, which are

typically recycled to reduce oil use and pollutant discharges. Subsequent operations may include drawing or extrusion to manufacture tube, wire, or die casting operations. In these operations, similar pollutants are discharged. Contaminated wet scrubber wastewaters may also be generated from extrusion processes, but to a lesser degree than in iron- and steelmaking and sintering operations.

* *Acid pickling:* Steel products are immersed in heated acid solutions to remove surface scale during pickling operations. This generates wastewater from three sources: (1) rinse water used to clean the product after immersion in pickling solution; (2) spent pickling solution or liquor; and (3) wastewater from wet fume scrubbers. The first source accounts for the largest volume of wastewater, but the second source is very acidic and contains high concentrations of iron and heavy metals.

* *Alkaline cleaning:* This process is used when vegetable, mineral, and animal fats and oils must be removed from the metal surface prior to further processing. Large-scale production or situations where a cleaner product is required may use electrolytic cleaning. The alkaline cleaning bath typically contains a solution of water, carbonates, alkaline silicates, phosphates, and sometimes wetting agents to aid cleaning. Alkaline cleaning results in the discharge of wastewaters from the cleaning solution tank and subsequent rinsing steps. Potential contaminants include dissolved metals, solids, and oils.

6.8.2 Effective BMPs

Primary metals manufacturing operations have experienced source reduction and recycle/reuse benefits similar to those available to metal-finishing operations including conserving waters through countercurrent rinsing techniques, and utilizing electrolytic recovery, customized resins, selective membranes, and adsorbents to separate metal impurities from acid/caustic dips and rinse waters to thereby allow for recycle and reuse.

Some very unique opportunities are also exclusively available to the primary metals industry. For example, the use of dry air control devices and dry cast quench operations have been adopted at some facilities to avoid the generation of contaminated wastewater. Additionally, many facilities are finding markets for by-products (e.g., sulfides resulting from nonferrous smelting operations can be converted to sulfuric acid and subsequently sold) which avoids the need to discharge these contaminants.

6.8.3 Case Studies

California Steel Industries, Inc., located in Fontana, California, reclaimed wastes to increase profits and address water use issues. The facility, a steel mill, is situated in an area that does not have a ready supply of process water. Also, the off-site recycling facility used to dispose of spent process pickle liquor was soon to become unavailable. As a result of these concerns, the company constructed an on-site recycling facility designed to recover ferrous chloride for resale and to reuse water and hydrogen chloride for use in steel-processing operations. Environmental benefits include the recovery and resale of 20 to 25 tons per day of ferrous chloride, 3,550 gallons per day of hydrogen chloride, and 13,000 gallons per day of water. In addition, corporate liability was minimized because spent liquor was no longer sent to a disposal facility.

6.9 PETROLEUM REFINING

The petroleum-refining industry uses chemical reactions and physical separation processes to create gasoline, residual fuel oil, jet fuel, heating oils and gases, petrochemicals, and a wide variety of other products from crude petroleum. Businesses classified as petroleum-refining facilities are represented by SIC group number 2911.

6.9.1 Industry Processes/Wastes Generated

A petroleum refinery is a complex combination of interdependent operations engaged in separating crude molecular constituents, molecular cracking, molecular rebuilding, and solvent blending and finishing to produce petroleum-derived products. More than 150 separate processes have been identified for the refining of crude petroleum and its products. Each unit of operation may be associated with quite different water usages. The types and quantities of contact wastewater produced are directly related to the nature of the various processes. Some major petroleum-refining processes and associated wastewater pollutants are described below.

* *Crude oil and product storage:* Crude oil, intermediate, and finished products are stored in tanks of varying sizes to provide adequate supplies for various refining processes. Operating schedules usually permit sufficient detention time for settling of water and suspended solids. Pollutants are mainly in the form of emulsified oil and suspended solids. Wastes are also a result of spills, leaks, and tank cleaning.

* *Ballast water storage:* Tankers which are used to ship intermediate and final products generally discharge ballast. Ballast waters are contaminated with product materials that are the crude feedstock in use at the refinery, ranging from water-soluble alcohol to residual fuels and brackish water.

* *Crude desalting:* Salts are separated from oil using emulsifiers and settling tanks. The wastewater stream from a desalter contains emulsified oil, ammonia, phenol, sulfides, suspended solids, and chlorides. Thermal pollution is also a problem in that the wastewater often exceeds 95^0 Celsius.

* *Crude oil fractionation:* Fractionation serves as the basic refining process for the separation of crude petroleum into intermediate fractions of specified boiling point ranges. Wastes include sour water drawn off from overhead accumulators prior to recirculation, which contains sulfides, ammonia, oil, chlorides, mercaptans, and phenols. Discharge from oil sampling lines also may contribute pollutants to wastewaters.

* *Catalytic cracking:* Catalytic cracking breaks heavy fractions such as oils into lower-molecular weight fractions. This process produces large volumes of high-octane gasoline stocks, furnace oils, and other middle-molecular weight distillates. Pollutants generally come from steam strippers

and overhead accumulators on fractioners. Major pollutants resulting from catalytic cracking are oil, phenols, sulfides, cyanides, ammonia, and carbon monoxide.

* *Solvent refining:* Solvents are used to extract contaminants from stock. The major pollutants from solvent refining are the solvents themselves. Under ideal conditions the solvents are continually recirculating with no losses to the sewer. Unfortunately, some solvent is always lost through pump seals, flange leaks, and elsewhere. Solvents are mostly lost from the bottom of fractionation towers and include phenol, glycol, and amines.

6.9.2 Effective BMPs

The petroleum-refining industry is unique in that its raw materials, wastes, and products are the same. Thus, source reduction measures such as materials substitution for crude oil are not realistic. Some facilities have, however, begun only accepting crude oil which meets certain quality specifications. Other facilities have implemented source reduction measures involving the use of less toxic catalysts and additives.

Recycle and reuse opportunities for the petroleum-refining operations are plentiful. Tank bottoms, slop oil, dissolved air flotation float, and American Petroleum Institute separator sludge are commonly recycled to the crude unit and, in some cases, the coker for further on-site processing to recover hydrocarbons. Many other by-products also can be recovered and reused on site. For example, recovered acids and caustics can be used for wastewater neutralization. Other recovered materials (i.e., spent catalyst) are often shipped off site for reuse in the paper industry or for further reclamation of precious metals such as vanadium. The proven use of BMPs by one industry is highlighted below.

6.9.3 Case Studies

A large petroleum-refining operation installed a Stretford Chemical Recovery Process unit to recover active sulfurs in wastewaters. By recovering these sulfurs, the frequency of solution dumping decreased from every $2\frac{1}{2}$ months to once per year, with a reduced disposal volume of 225,000 to 25,000 gallons per year. Thus, releases of sulfur to the wastewater and the need for dumping and off-site disposal of sulfur were also reduced. This helped minimize risks of soil contamination at disposal sites. Ultimately, the savings including $60,000 per year in disposal costs and $120,000 per year in raw materials resulted.

6.10 INORGANIC CHEMICALS MANUFACTURING

The inorganic chemicals industry is very large and diversified. Thirty-five major categories of inorganic chemicals are known to generate polluted wastewater. The major industries include manufacturers of alkalies and chlorine (SIC group number 2812), industrial gases (SIC group number 2813), inorganic pigments (SIC group number 2816), and industrial inorganic chemicals not elsewhere classified (SIC group number 2819).

6.10.1 Industry Processes/Wastes Generated

Inorganic chemicals are manufactured for captive or merchant use in four or more steps moving from raw material to final product. Two or more different products may use the same process, but the raw materials used, process sequence, control, recycle potential, handling, and quality control vary among products, as does the quality of wastes.

Plant process wastewaters from the inorganic chemicals industry often contain toxic metals such as mercury, zinc, chromium, lead, arsenic, cadmium, nickel, silver, copper, and cyanide. Very few organic toxic pollutants are found in process waste streams, and those found tend to be present in low-level concentrations. All of the processes tend to have discharges of acids such as sulfuric, hydrofluoric, and hydrochloric. Other substances often present in the wastewater are salts, asbestos, total residual chlorine, and iron.

6.10.2 Effective BMPs

BMPs directed at reducing wastewater consumption have greatly reduced costs in the inorganic chemical manufacturing industry. While treatment technologies have been most often used to reduce the levels of toxic metals and other pollutants in the wastewater, source reduction and recovery/reuse practices have also been employed.

Many of the most effective pollution prevention mechanisms in the inorganic chemicals manufacturing industry have employed the reclamation of by-products, which had previously been considered wastes, and the development of saleable products. Additionally, as with the OCPSF manufacturing industry, the inorganics-manufacturing industry has found success in substituting less-toxic catalysts and in exhibiting better reaction controls, thus minimizing or eliminating excess toxic and hazardous wastes.

Chapter 7

ACTIVITY-SPECIFIC BMPs

7.1 INTRODUCTION

This chapter describes specific Best Management Practices (BMPs) for common industrial activities that may contaminate storm water. The suggested BMPs are relatively easy to use, are inexpensive, and often are effective in removing the source of storm water contaminants. This chapter does not present a complete listing of BMPs for every industrial activity; rather, it is meant to help facility owners and operators implement different methods for reducing storm water contamination at the facility site.

7.2 FUELING STATION ACTIVITIES

When storm water mixes with fuel spilled or leaked onto the ground, it becomes polluted with chemicals that are harmful to human health and to fish and wildlife. This section identifies fueling station activities that can contaminate storm water and suggests BMPs to help reduce or eliminate storm water contamination from fueling stations. Common fuel station activities that can contaminate storm water include:

* Spills and leaks that happen during fuel or oil delivery.

* Spills caused by "topping off" fuel tanks.

* Allowing rainfall on the fuel area or storm water to run onto the fuel area.

* Housing or washing down the fuel area.

* Leaking storage tanks.

7.2.1 Overfill Prevention Equipment

Fuel overflows during storage tank filling are a major source of spills. Overflows can be prevented. Watch the transfer constantly to prevent overfilling and spilling. Overfill prevention equipment automatically shuts off flow, restricts flow, or sounds an alarm when the tank is almost full.

Resource Conservation and Recovery Act (RCRA) regulations require overfill prevention equipment on all underground storage tanks (USTs) installed after December 1988. For USTs installed before December 1988, overfill prevention equipment is required by 1998. However, state and local regulations may be more stringent, so contact state and/or local authorities for details. Consider installing overflow prevention equipment sooner than required as part of your pollution prevention plan.

7.2.2 Discourage "Topping Off" Vehicle Fuel Tanks

Gas pumps automatically shut off when the vehicle fuel tank is almost full to prevent spills. Trying to completely fill the tank often results in overfilling the tank and spilling fuel. Discourage "topping off" by training employees and posting signs.

7.2.3 Protect Fuel Areas from Rain

Fueling areas can be designed to minimize spills, leaks, and incidental losses of fuel, such as vapor loss, from coming into contact with rain water through implementation of the following practices:

* Build a roof over the fuel area.

* Pave the fuel area with concrete instead of asphalt. Asphalt soaks up fuel or can be slowly dissolved by fuel, engine fluids, and other organic liquids. Over time, the asphalt itself can become a source of storm water contamination.

7.2.4 Minimize Runon to the Fueling Area

Runon is storm water generated from other areas that flows or "runs on" to the site. Runon flowing across fueling areas can wash contamination into storm drains. The following practices can be used to minimize runon:

* Grading, berming, or curbing the area around the fuel site to direct runon away from the fuel area.

* Locating roof downspouts so storm water is directed away from fueling areas.

* Using valley gutters to route storm water around fueling areas.

7.2.5 Install Oil and Grease Traps

Oil/water separators and oil and grease traps are devices that reduce the amount of oil entering storm drains. These devices should be installed and routinely inspected, cleaned, and maintained.

7.2.6 Avoid Hosing or Washing Down Fuel Areas

Cleaning the fueling area with running water should be avoided because the wash water will pick up fuel, oil, and grease and become contaminated storm water. Consider using a damp cloth on the fuel pumps and a damp mop on the pavement rather than a hose. Check with the local sewer authority about any treatment required before discharging the mop water or wash water to sanitary sewers.

7.2.7 Spill Control Measures

Spills should be controlled immediately. Small spills can be contained using a sorbent material, such as cat litter, straw, or sawdust. Do not wash petroleum spills into the storm drain or sanitary sewer. (see Chapter 8 for further information on spill control measures, such as containment diking and curbing.)

7.2.8 Encourage Employee Participation

Storm water contamination from fueling operations often occurs from small actions such as topping off fuel tanks, dripping engine fluids, and hosing down fuel areas. Inform employees about ways to eliminate or reduce storm water contamination.

Getting employees involved in waste minimization is a key to a successful Storm Water Pollution Prevention Plan. Because employees are most familiar with operations that generate wastes, they may have helpful waste reduction suggestions. Consider setting up an employee reward program to promote pollution prevention.

7.3 MAINTAINING VEHICLES AND EQUIPMENT

Many vehicle- and equipment-maintenance operations use materials or create wastes that are harmful to humans and the environment. Storm water runoff from areas where these activities occur can become polluted by a variety of contaminants such as solvents and degreasing products, waste automotive fluids, oils and greases, acids, and caustic wastes. These and other harmful substances in storm water can enter water bodies through storm drains or through small streams where they can harm fish and wildlife. The following are vehicle- and equipment-maintenance activities that can contaminate storm water:

* Engine repair and service:
  - parts cleaning
  - shop cleaning
  - spilled fuel, oil, or other materials
  - replacement of fluids (oil, oil filters, hydraulic fluids, transmission fluid, and radiator fluids)

* Outdoor vehicle and equipment storage and parking:
  - dripping engine and automotive fluids from parked vehicles and equipment

* Disposal of materials or process wastes:
  - greasy rags
  - oil filters
  - air filters
  - batteries
  - spent coolant, degreasers, etc.

7.3.1 Parts Cleaning

Parts are often cleaned using solvents such as trichloroethylene, 1,1,1-trichloroethane, or methylene chloride. Many of these cleaners are harmful and must be disposed of as hazardous wastes. Cleaning without using liquid cleaners whenever possible reduces waste. Scrape parts with a wire brush, or use a bake oven if one is available. Prevent spills and drips of solvents and cleansers to the shop floor. Do all liquid cleaning at a centralized station so the solvents and residues stay in one area. If parts are dipped in liquid, remove them slowly to avoid spills. Locate drip pans, drain boards, and drying racks to direct drips back into a sink or fluid holding tank for reuse.

7.3.2 Substitute Nontoxic or Less Toxic Cleaners

If possible, eliminate or reduce the number or amount of hazardous materials and waste by substituting nonhazardous or less hazardous materials. For example:

* Use noncaustic detergents instead of caustic cleaning agents for parts cleaning (ask your supplier about alternative cleaning agents).

* Use detergent-based or water-based cleaning systems in place of organic solvent degreasers. Wash water may require treatment before it can be discharged to the sanitary sewer. Contact your local sewer authority for more information.

* Replace chlorinated organic solvents (1,1,1-trichloroethane, methylene chloride, etc.) with nonchlorinated solvents. Nonchlorinated solvents like kerosene or mineral spirits are less toxic and less expensive to dispose of, but are by no means harmless themselves. Check the list of active ingredients to see whether it contains chlorinated solvents.

* Choose cleaning agents that can be recycled.

* Contact your supplier or trade journal for more waste minimization ideas.

7.3.3 Use Dry Cleanup Methods

Clean up leaks, drips, and other spills without large amounts of water. Use rags for small spills, a damp mop for general cleanup, and dry absorbent material for larger spills. Consider using the following BMPs:

* Avoid hosing down work areas.

* Collect leaking or dripping fluids in drip pans or containers. If different liquids are kept separate, the fluids are easier to recycle.

* Keep a drip pan under the vehicle while unclipping hoses, unscrewing filters, or removing other parts. Use a drip pan under any vehicle that might leak while working on it to keep splatters or drips off the shop floor.

* Promptly transfer used fluids to the proper waste or recycling drums. Don't leave full drip pans or other open containers lying around.

* Locate waste and recycling drums in properly controlled areas of the yard, preferably areas with a concrete slab and secondary containment.

7.3.4 Prevent Pouring of Liquid Wastes Down Drains

Do not pour liquid waste to floor drains, sinks, outdoor storm drain inlets, or other storm drains or sewer connections. Used or leftover cleaning solutions, solvents, and automotive fluids and oil are often toxic and should not be put into the sanitary sewer. Be sure to dispose of these materials properly or find opportunities for reuse and recycling. If unsure of how to dispose of chemical wastes, contact your State hazardous waste management agency or the RCRA hotline at 1-800-424-9346. Post signs at sinks to remind employees, and paint stencils at outdoor drains to tell customers and others not to pour wastes down drains.

7.3.5 Drain Oil Filters Before Disposal

Oil filters disposed of in trash cans or dumpsters can leak oil and contaminate storm water. Place the oil filter in a funnel over the waste oil recycling or disposal collection tank to drain excess oil before disposal. Oil filters can be crushed and recycled. Ask your oil supplier or recycler about recycling oil filters.

7.3.6 Check Incoming Vehicles and Equipment for Leaks

If possible, park vehicles indoors or under a roof so storm water does not contact the area. If vehicles are parked outdoors while they await repair, watch them closely for leaks.

Put pans under leaks to collect fluids for proper recycling or disposal. Keeping leaks off the ground reduces the potential for storm water contamination and reduces cleanup time and costs. If the vehicle or equipment is to be stored outdoors, oil and other fluids should be drained first.

Designate a special area to drain and replace motor oil, coolant, and other fluids, where there are no connections to the storm drain or the sanitary sewer and drips and spills can be easily cleaned up.

7.3.7 Exercise Care with Wrecked Vehicles and Equipment

Be especially careful with wrecked vehicles, whether kept indoors or outside, as well as with vehicles kept on site for scrap or salvage. Wrecked or damaged vehicles often drip oil and other fluids for several days. The following practices can help reduce storm water contamination from wrecked vehicles and equipment:

* As the vehicles arrive, place drip pans under them immediately, even if you believe that all fluids have leaked out before the car reaches your shop.

* Build a shed or temporary roof over areas where you park cars awaiting repairs or salvage, especially if you handle wrecked vehicles. Build a roof over vehicles kept for parts.

* Drain all fluids, including air conditioner coolant, from wrecked vehicles and "parts" cars. Also drain engines, transmissions, and other used parts.

7.3.8 Handling Cracked Batteries

Store cracked batteries in a nonleaking secondary container. Do this with all cracked batteries, even if you think all the acid has drained out. If you drop a battery, treat it as if it is cracked. Put it into the containment area until you are sure it is not leaking.

Handle spilled acid from broken batteries with care. If baking soda is used to neutralize spilled battery acid during cleanup, remember that residue is still dangerous to handle and must be disposed of as a hazardous waste because it may contain lead and other contaminants.

7.3.9 Recycle Engine and Equipment Fluids

Various engine and equipment fluids, including degreasers, used oil or oil filters, antifreeze, cleaning solvents, and hydraulic fluids can either be recycled on site or sent off site for recycling. The following are some recycling options, ranked by level of effort required:

* *Least Effort:*

  - Arrange for collection and transportation of car batteries, used oil and other fluids, cleaning solutions, and degreasers to a commercial recycling facility. This requires separation of wastes and storage until they are picked up by the recycling company.

- "Dirty" solvent can be reused. Presoak dirty parts in used solvent before cleaning the parts in fresh solvent.

* *Moderate Effort:*

- Used oil, antifreeze, and cleaning solutions can be recycled on site using a filtration system that removes impurities and allows the fluid to be reused. Filtration systems are commercially available.

* *Most Effort:*

- Install an on-site solvent-recovery unit. If your facility creates large volumes of used solvents, consider purchasing or leasing an on-site still to recover the solvent for reuse. Contact the state hazardous waste management agency for more information about on-site recycling of used solvents.

Reducing the number of solvents makes recycling easier and reduces hazardous waste management costs. Often, one solvent can perform a job as well as two different solvents.

Separating wastes also allows for easier recycling and may reduce treatment costs. Keep hazardous and nonhazardous wastes separate, do not mix used oil and solvents, and keep chlorinated solvents (like 1,1,1-trichloroethane) separate from nonchlorinated solvents (like kerosene and mineral spirits). Proper labeling of all wastes and materials will help accomplish this goal.

Finally, many products made of recycled (i.e., refined or purified) materials are available. Engine oil, transmission fluid, antifreeze, and hydraulic fluid are available in recycled form. Buying recycled products supports the market for recycled materials.

7.4 PAINTING VEHICLES AND EQUIPMENT

Many painting operations use materials or create wastes that are harmful to humans and the environment. Storm water runoff from areas where these activities occur can become polluted by a variety of contaminants such as solvents and dusts from sanding and grinding that contain toxic metals like cadmium and mercury. These and other potentially harmful substances in storm water can enter water bodies directly through storm drains where they can harm fish and wildlife.

This section identifies potential sources of storm water contamination from painting operations and BMPs that can reduce or eliminate these sources. The guidance provided in this section can be used to eliminate, reduce, or recycle pollutants that may otherwise contaminate storm water. The following painting activities can contaminate storm water:

* Painting and paint removal.

* Sanding or paint stripping.

* Spilled paint or paint thinner.

7.4.1 Use Spill Collection Devices

Use tarps and vacuums to collect solid wastes produced by sanding or painting. Tarps, drip pans, or other spill collection devices should be used to collect spills of paints, solvents, or other liquid materials. These wastes should be disposed of properly to keep them from contaminating storm water.

7.4.2 Contain Sanding Wastes

Prevent paint chips from coming into contact with storm water. Paint chips may contain hazardous metallic pigments or biocides. You can reduce contamination of storm water with paint dust and chips from sanding by using the following practices:

* Avoid sanding in windy weather when possible.

* Enclose outdoor sanding areas with tarps or plastic sheeting. Be sure to provide adequate ventilation and personal safety equipment. After sanding is complete, collect the waste and dispose of it properly.

* Keep workshops clean of debris and grit so that the wind will not carry any waste into areas where it can contaminate storm water.

* Move the activity indoors if you can do so safely.

7.4.3 Use Proper Paint Spray Equipment/Techniques

As little as 30% of the paint may reach the target from conventional airless spray guns; the rest is lost as overspray. Paint solids from overspray are deposited on the ground where they can contaminate storm water. Other spray equipment that delivers more paint to the target and less overspray should be used, such as:

* Electrostatic spray equipment.

* Air-atomized spray guns.

* High-volume/low-pressure spray guns.

* Gravity-feed guns.

Operator training can also reduce overspray and minimize the amount of paint solids that can contaminate storm water. Correct spraying techniques also reduce the amount of paint needed per job. If possible, avoid spraying on windy days. When spraying outdoors, use a drop cloth or ground cloth to collect and dispose of overspray.

7.4.4 Recycle Paint, Paint Thinner, or Solvents

Paint, paint thinner, and solvents can either be recycled at the facility or sent off site for recycling. The following are some recycling options ranked by the level of effort required:

* *Least Effort:*

  - Dirty solvent can be reused for cleaning dirty spray equipment and parts before equipment is cleaned in fresh solvent.

- Give small amounts of leftover paint to the customer for touchup.

* *Moderate Effort:*

- Arrange for collection and transportation of paints, paint thinner, or spent solvents to a commercial recycling facility.

* *Most Effort:*

- Install an on-site solvent recovery unit. If your facility creates large volumes of used solvents, paint, or paint thinner, you may consider buying or leasing an on-site still to recover used solvent for reuse. Contact the state hazardous waste management agency for more information about on-site recycling of used solvents.

Separating wastes makes recycling easier and may reduce treatment costs. Keep hazardous and nonhazardous wastes separate, and keep chlorinated solvents (like 1,1,1-trichloroethane) separate from nonchlorinated solvents (like petroleum distillate and mineral spirits). Check the materials data sheet for ingredients, or talk with your waste hauler or recycling company to learn which waste types can be stored together and which should be separated.

Reducing the number of solvents also makes recycling easier and reduces hazardous waste management costs. Often, one solvent can do a job as well as two different solvents.

Finally, many products made of recycled (i.e., refined or purified) materials are available. Buying recycled paints, paint thinner, or solvent products helps build the market for recycled materials.

7.5 WASHING VEHICLES AND EQUIPMENT

Washing vehicles and equipment outdoors or in areas where wash water flows onto the ground can pollute storm water. Wash water can contain high concentrations of oil and grease, phosphates, and high suspended solid loads (these and other potentially harmful substances can pollute storm water when deposited on the ground where they can be picked up by rainfall runoff). Vehicle wash water is considered to be a process wastewater and needs to be covered by a National Pollutant Discharge Elimination System (NPDES) permit. Contact the permitting authority for information about how vehicle wash water is being regulated.

This section identifies sources of storm water contamination from vehicle and equipment washing and offers BMPs to reduce those sources. The guidance provided in this section can be used to eliminate, reduce, or recycle pollutants that otherwise may contaminate storm water. The following vehicle and equipment washing activities can contaminate storm water:

* Outside equipment or vehicle cleaning (washing or steam cleaning).

* Wash water discharged directly to the ground or storm water drain.

7.5.1 Use Phosphate-Free Detergents

Phosphates, which are plant nutrients, can cause excessive growth of nuisance plants in water when they enter lakes or streams in wash water. Some states ban the use of detergents containing high amounts of phosphates. Contact suppliers about phosphate-free biodegradable detergents that are available on the market.

7.5.2 Avoid Outdoor Washing

Used wash water contains high concentrations of solvents, oil and grease, detergents, and metals. Try not to wash parts or equipment outside. Washing over impervious surfaces like concrete, blacktop, or hardpacked dirt allows wash water to enter storm drains directly or deposits contaminants on the ground, where they are washed into storm drains when it rains. Washing over pervious ground, such as sandy soils, potentially can pollute ground water. Therefore, small parts and equipment washing should be done over a parts-washing container where the wash water can be collected and recycled or disposed of properly.

7.5.3 Use Designated Cleaning Areas

If you are washing large equipment or vehicles, and have to wash outside, designate a specific area for washing. This area should be bermed to collect the wastewater and graded to direct the wash water to a treatment facility. Consider filtering and recycling vehicle wash water. If recycling is not practical, the wastewater can be discharged to the sanitary sewer. Contact the local sewer authority to find out whether treatment is required before wash water is discharged to the sewer (pretreatment).

7.6 LOADING AND UNLOADING VEHICLES

Loading/unloading operations usually take place outside on docks or terminals. Materials spilled, leaked, or lost during loading/unloading may collect in the soil or on other surfaces and be carried away by rainfall runoff or when the area is cleaned. Rainfall may wash off pollutants from machinery used to unload or load materials. This section identifies sources of storm water contamination from loading and unloading materials and offers BMPs to reduce or eliminate those sources. The guidance provided in this section can be used to eliminate, reduce, or recycle pollutants that otherwise may contaminate storm water. The following loading and unloading activities can contaminate storm water:

* Pumping of liquids or gases from barge, truck, or rail car to a storage facility or vice versa.

* Pneumatic transfer of dry chemicals to or from the loading and unloading vehicles.

* Transfer by mechanical conveyor systems.

* Transfer of bags, boxes, drums, or other containers by forklift, trucks, or other material-handling equipment.

7.6.1 Check Vehicles/Equipment for Leaks

Check vehicles and equipment regularly for leaks, and fix any leaks promptly. Common areas for leaks are valves, pumps, flanges, and connections. Look for dust or fumes. These are signs that material is being lost during unloading/loading operations.

Loading/unloading equipment and vehicles should be located so that leaks can be contained in existing containment and flow diversion systems.

7.6.2 Prevent Storm Water Runon

Runon is storm water created from other areas that flows or "runs on" to the site. Runon flowing across loading/unloading areas can wash contaminants into storm drains. Runon can be minimized by:

* Grading, berming, or curbing around the loading area to direct runon away from the area.

* Positioning roof downspouts so storm water is directed away from loading sites and equipment and preferably to a grassy or vegetated area where the storm water can soak into the ground.

Finally, by covering loading and unloading areas, such as building overhangs at loading docks, exposure of materials, vehicles, and equipment to rain can be reduced.

7.7 LIQUID STORAGE IN ABOVEGROUND TANKS

Accidental releases of chemicals from aboveground liquid storage tanks can contaminate storm water with many different pollutants. Materials spilled, leaked, or lost from storage tanks may accumulate in soils or on other surfaces and be carried away by rainfall runoff. This section identifies sources of storm water contamination from aboveground storage tanks and offers BMPs to reduce or eliminate those sources. The following is a list of the most common causes of unintentional releases from aboveground tanks:

* External corrosion and structural failure.

* Installation problems.

* Spills and overfills due to operator error.

* Failure of piping systems (pipes, pumps, flanges, couplings, hoses, and valves).

* Leaks or spills during pumping of fluids or gases from barges, trucks, or rail cars to a storage facility or vice versa.

7.7.1 Compliance with Regulatory Standards

Storage of oil and hazardous materials must meet specific standards set by Federal and State laws. These standards include Spill Prevention Control and Countermeasures (SPCC) plans, secondary containment, installation, integrity and leak detection monitoring, and emergency preparedness plans. Federal regulations set specific standards for preventing runon and collecting runoff from hazardous waste storage, disposal, or treatment areas. These standards apply to container storage areas and other areas used to store, treat, or dispose of hazardous waste. If the collected storm water is a hazardous waste, it must be managed as a hazardous waste in accordance with all applicable State and Federal environmental regulations. States may also have standards about controlling runon and runoff from hazardous waste treatment, storage, and disposal areas. To find out more about storage requirements, call the EPA's toll-free RCRA hotline at 1-800-424-9346 or contact the State hazardous waste management agency.

7.7.2 Safeguards Against Accidental Releases

Well-trained employees can reduce human errors that lead to accidental releases or spills. In addition, engineering safeguards can help prevent operator errors that may cause the accidental release of pollutants. Such safeguards include:

* Overflow protection devices on tank systems to warn the operator or to automatically shut down transfer pumps when the tank reaches full capacity.

* Protective guards around tanks and piping to prevent vehicle or forklift damage.

* Clearly tagging or labeling valves to reduce human error.

7.7.3 Routinely Inspect Tanks and Equipment

Visually inspect the tank system to identify problem areas before they lead to a release. Correct any problems or potential problems as soon as possible. An audit of a newly installed tank system by a registered and specially trained professional engineer can identify and correct potential problems such as loose fittings, poor welding, and improper or poorly fitted gaskets. After installation, have operators visually inspect the tank system on a routine basis. Areas to inspect include tank foundations, connections, coatings, tank walls, and the piping system. Look for corrosion, leaks, straining of tank support structures from leaks, cracks, scratches in protective coatings, or other physical damage that may weaken the tank system. Integrity testing should be done periodically by a qualified professional.

7.7.4 Install Secondary Containment

A secondary containment system around both permanent and temporary tanks allows leaks to be more easily detected and contains spills or leaks. Methods include berms, dikes, liners, vaults, and double-walled tanks. (see Chapter 8 for additional information on containment and spill control.)

7.8 INDUSTRIAL WASTE MANAGEMENT AND OUTSIDE MANUFACTURING

Storm water runoff from areas where industrial waste is stored, treated, or disposed of can be polluted. Outside manufacturing activities can also contaminate storm water runoff. Activities such as rock grinding or crushing, painting or coating, grinding or sanding, degreasing or parts cleaning, or operations that use hazardous materials are particularly dangerous. Wastes spilled, leaked, or lost from waste management areas or outside manufacturing activities may build-up in soils or on other surfaces and be carried away by rainfall runoff. There is also a potential for liquid wastes from lagoons or surface impoundments to overflow to surface waters or soak the soil where they can be picked up by storm water runoff. Possible storm water contaminants include toxic compounds, oil and grease, paints or solvents, heavy metals, and high levels of suspended solids. The following industrial waste management activities or areas can contaminate storm water:

* Waste piles.

* Wastewater and solid waste treatment and disposal:
  - Waste pumping
  - Additions of treatment chemicals
  - Mixing
  - Aeration
  - Clarification
  - Solids dewatering

* Land application.

In addition, the following outside manufacturing activities or situations can contaminate storm water:

* Processes or equipment that generate dusts, vapors, or emissions.

* Outside storage of hazardous materials or raw materials.

* Dripping or leaking fluids from equipment or processes.

* Liquid wastes discharged directly onto the ground or into the storm sewer.

The best way to reduce the potential for storm water contamination from both waste management areas and outside manufacturing activities is to reduce the amount of waste that is created and, consequently, the amount that must be stored or treated. The following section offers BMPs that can eliminate or reduce the amount or toxicity of industrial wastes as well as minimize contamination of storm water from existing waste management areas. Waste reduction BMPs are appropriate for a wide range of industries and are designed to provide ideas on ways to reduce wastes.

7.8.1 Conduct a Waste Reduction Assessment

The first step to reducing wastes is to assess activities at the facility. The assessment is designed to find situations at the facility where waste generation, emissions, and environmental damage can be eliminated or reduced. The assessment involves steps very similar to those used to develop a Storm Water Pollution Prevention Plan, such as collecting process-specific information; setting pollution prevention targets; and developing, screening, and selecting waste reduction options for further study. Starting a waste reduction program at the facility has many potential benefits. Some of these benefits are direct (e.g., cost savings from reduced raw material use), while others are indirect (e.g., avoided waste disposal fees).

There are many different types of BMPs that can help eliminate or reduce the amount of industrial waste generated at the facility. Some of these BMPs are listed below:

* Production planning and sequencing.
* Process or equipment modification.
* Raw material substitution or elimination.
* Loss prevention and housekeeping.
* Waste segregation and separation.
* Closed-loop recycling.
* Training and supervision.
* Reuse and recycling.

7.8.2 Check for Spills and Leaking Containers

By checking waste management areas for leaking containers or spills, wastes can be prevented from contaminating storm water. Look for containers that are rusty, corroded, or damaged. Transfer wastes from these damaged containers into safe containers. Close the lids on dumpsters to prevent rain from washing wastes out of holes or cracks in the bottom of the dumpster.

In outside manufacturing areas, look for leaking equipment (e.g., valves, lines, seals, or pumps) and fix leaks promptly. Inspect rooftop and other outdoor equipment (e.g., HVAC devices, air pollution control devices, transformers, piping, etc.) for leaks or dust concentrations.

7.8.3 Prevent Runon or Rain from Contacting Waste Management Areas

The best way to avoid contaminating storm water from existing waste management and manufacturing areas is to prevent storm water runon or rain from entering or contacting these areas. This can be done by:

* Preventing direct contact with rain.

* Moving the activity indoors after ensuring that all safety concerns, such as fire hazard and ventilation, are addressed.

* Covering the area with a permanent roof.

* Covering waste piles with a temporary covering material, such as a reinforced tarpaulin, polyethylene, polyurethane, polypropylene, or Hypalon.

* Minimizing storm water runon by enclosing the area or building a berm around the area.

7.8.4 Furnish Transport Vehicles with Spill Prevention Equipment

Transport vehicles equipped with spill prevention equipment can prevent spills of wastes during transport. Examples include:

* Vehicles equipped with baffles for liquid wastes.

* Trucks with sealed gates and spill guards for solid wastes.

* Trucks with tarps.

7.8.5 Minimize Runoff from Land Application Sites

Certain precautions can be taken to minimize the runoff of polluted storm water from land application sites. Some of these precautions include:

* Choose the land application site carefully. Characteristics that help prevent runoff include slopes under 6%, permeable soils, a low water table, locations away from wetlands or marshes, and closed drainage systems.

* Avoid applying waste to the site when it is raining or when the ground is frozen or saturated with water. Grow vegetation on areas dedicated to land disposal to stabilize the soils and reduce the volume of surface water runoff from the site.

* Maintain adequate barriers between the land application site and receiving waters.

* Erosion control techniques might include mulching and matting, filter fences, straw bales, diversion terracing, or sediment basins (see Chapter 8 for a detailed description of erosion control techniques).

* Perform routine maintenance to ensure that erosion control or site stabilization measures are working.

7.9 OUTSIDE STORAGE OF RAW MATERIALS, BY-PRODUCTS, OR FINISHED PRODUCTS

Raw materials, by-products, finished products, containers, and material storage areas exposed to rain and/or runoff can pollute storm water. Storm water can become contaminated by a wide range of contaminants (e.g., metals, oil, and grease) when solid materials wash off or dissolve into water, or by spills or leaks. This section identifies potential sources of storm water contamination and offers BMPs that can reduce or eliminate those sources.

The best way to avoid contaminating storm water from outside material storage areas is to prevent storm water runon or rain from coming in contact with the materials. This can be done by:

* Storing the material indoors.

* Covering the area with a roof.

* Covering the material with a temporary covering made of polyethylene, polyurethane, polypropylene, or Hypalon.

* Minimizing storm water runon by enclosing the area or building a berm around the area.

7.10 SALT STORAGE FACILITIES

Salt left exposed to rain or snow can be lost. Salt spilled or blown onto the ground during loading and unloading will dissolve in storm water runoff. Storm water contaminated with salt can be harmful to vegetation and aquatic life. Salty storm water runoff soaking into the ground may contaminate ground water and make it unsuitable as a drinking water supply. The following salt storage activities can contaminate storm water:

* Salt stored outside in piles or bags that are exposed to rain or snow.

* Salt loading and unloading areas located outside or in areas where spilled salt can contaminate storm water.

This section offers BMPs to help reduce storm water contamination from salt storage and transfer activities. The best way to prevent salt from contaminating storm water is to eliminate or limit the exposure of salt to rain. Store salt under a roof. This is the best way to stop direct contact with rain. Preventing contact with rain also protects against salt loss and prevents salt from absorbing moisture and becoming caked or lumpy and making it difficult to handle and use. Storm water runon can also be minimized by building berms around storage, loading, and unloading areas.

If salt must be stored outside:

* Build the storage pile on concrete or asphalt to reduce the potential for groundwater contamination.

* Cover the pile with a temporary covering material, such as polyethylene, polyurethane, polypropylene, or Hypalon.

Chapter 8

FLOW DIVERSION, EXPOSURE MINIMIZATION, AND MITIGATION PRACTICES

8.1 INTRODUCTION

This chapter describes site-specific Best Management Practices (BMPs) designed to minimize, reduce, and eliminate storm water contamination from various industrial activities. Several different flow diversion, exposure minimization, and mitigation practices are provided to help in reducing storm water contamination at industrial facility sites.

8.2 FLOW DIVERSION PRACTICES

Structures that divert stream flow (such as gutters, drains, sewers, dikes, and graded pavement) are used as BMPs in two ways. First, flow diversion structures, called storm water conveyances, may be used to channel storm water away from industrial areas so that pollutants do not mix with the storm water. Second, they also may be used to carry pollutants directly to a treatment facility. This section briefly describes flow diversion as a BMP for industrial storm water.

8.2.1 Storm Water Conveyances

Storm water conveyances (such as channels, gutters, drains, and sewers) collect storm water runoff and direct its flow. A group of connecting conveyances is sometimes installed at an industrial facility to create a storm water collection system. Storm water conveyances can be used for two different purposes. The first purpose is to keep uncontaminated storm water from coming in contact with areas of an industrial site where it may become contaminated with pollutants. This can be accomplished by collecting the storm water in a conveyance and by changing the direction of flow away from those areas. The second purpose is to collect and carry the storm water that has already come into contact with industrial areas and become contaminated to a treatment facility.

Storm water conveyances can be constructed or lined with many different materials, including concrete, clay tiles, asphalt, plastics, metals, riprap, compacted soils, and vegetation. The type of material used depends on the use of the conveyance. These conveyances can be temporary or permanent.

Storm water conveyances work well at most industrial sites. Storm water can be directed away from industrial areas by collecting it in channels or drains before it reaches the areas. In addition, conveyances can be used to collect storm water downhill from industrial areas and keep it separate from runoff that has been in contact with those areas. When potentially contaminated storm water is collected in a conveyance like this, it can be directed to a treatment facility on the site if necessary. (If a pollutant is spilled, it should not be allowed to enter a storm water conveyance or drain system.)

In planning for storm water conveyances, consider the amount and speed of the typical storm water runoff. Also, consider the patterns in which the storm water drains so that the channels may be located to collect the most flow and can be built to handle the amount of water they will receive. When deciding on the type of material for the conveyance, consider the resistance of the material, its durability, and compatibility with any pollutants it may carry.

Conveyance systems are most easily installed when a facility is first being constructed. Use of existing grades will decrease costs. Grades should be positive to allow for the continued movement of the runoff through the conveyance system; however, grades should not create an increase in velocity that causes an increase in erosion (this will also depend upon what materials the conveyance is lined with and the types of outlet controls that are provided).

Ideally, storm water conveyances should be inspected to remove debris within 24 hours of rainfall, or daily during periods of prolonged rainfall, since heavy storms may clog or damage them. It is important to repair damages to these structures as soon as possible.

The advantages and disadvantages of storm water conveyances are presented below:

Advantages of Storm Water Conveyances
(Channels / Gutters / Drains / Sewers)

* Direct storm water flows around industrial areas.

* Prevent temporary flooding of industrial sites.

* Require low maintenance.

* Provide erosion-resistant conveyance of storm water runoff.

* Provide long-term control of storm water flows

Disadvantages of Storm Water Conveyances
(Channels / Gutters / Drains / Sewers)

* Once flows are concentrated in storm water conveyances, they must be routed through stabilized structures all the way to their discharge to the receiving water or treatment plant to minimize erosion.

* May increase flow rates.

* May be impractical if there are space limitations.

* May not be economical, especially for small facilities or after a site has already been constructed.

8.2.2 Diversion Dikes

Diversion dikes or berms are structures used to block runoff from passing beyond a certain point. Temporary dikes are usually made with compacted soil. More permanent ridges are constructed out of concrete, asphalt, or similar materials.

Diversion dikes are used to prevent the flow of storm water runoff onto industrial areas. Limiting the volume of flow across industrial areas reduces the volume of storm water that may carry pollutants from the area, requiring treatment for pollutant removal. This BMP is suitable for industrial sites where significant volumes of storm water runoff tend to flow onto active industrial areas. Typically, dikes are built on slopes just uphill from an industrial area together with some sort of a conveyance such as a swale. The storm water conveyance is necessary to direct the water away from the dike so that the water will not pool and seep through the dike.

In planning for the installation of dikes, consider the slope of the drainage area, the height of the dike, the size of rainfall event it will need to divert, and the type of conveyance that will be used with the dike. Steeper slopes result in higher volumes of runoff and higher velocities; therefore, the dike must be constructed to handle this situation. Remember that dikes are limited in their ability to manage large volumes of runoff.

Ideally, dikes are installed before industrial activity begins. However, dikes can be easily constructed at any time. Temporary dikes (usually made of dirt) generally only last for 18 months or less, but they can be made into permanent structures by stabilizing them with vegetation. Vegetation is crucial for preventing the erosion of the dike.

Dikes should be inspected regularly for damage. This is especially important after storm events, since a heavy rain may wash parts of a temporary dike away. Any necessary repairs should be made immediately to make sure the structure continues to do its job.

The advantages and disadvantages of diversion dikes are presented below:

Advantages of Diversion Dikes:

* Effectively limit storm water flows over industrial site areas.

* Can be installed at any time.

* Are economical temporary structures when built from soil on site.

* Can be converted from temporary to permanent at any time.

Disadvantages of Diversion Dikes:

* Are not suitable for large drainage areas unless there is a gentle slope.

* May require maintenance after heavy rains.

8.2.3 Graded Areas and Pavement

Land surfaces can be graded or graded and paved so that storm water runoff is directed away from industrial activity areas. The slope of the grade allows the runoff to flow, but limits the runoff from washing over areas that may be contaminated with pollutants. Like conveyances and dikes, graded areas can prevent runoff from contacting industrial areas and becoming contaminated with pollutants from these areas. Grading can be a permanent or temporary control measure.

Grading land surfaces is appropriate for any industrial site that has outdoor activities that may contaminate storm water runoff, such as parking lots or outdoor storage areas. Grading is often used with other practices, such as coverings, buffer zones, and other practices to reduce the runoff velocity and provide infiltration of the uncontaminated runoff, or to direct pollutant runoff to storm water treatment facilities.

When designing graded areas and pavement, both control and containment of runoff flows should be considered. The grading should control the uncontaminated flow by diverting it around areas that may have pollutants. The grading should also contain the contaminated flows or divert them to treatment facilities.

When regrading and paving an industrial area, the use of concrete paving instead of asphalt should be considered. This is especially important in potential spill sites or hazardous material storage areas. Asphalt absorbs organic pollutants and can be slowly dissolved by some fluids, thus becoming a possible source of contaminants itself. This control measure should be used with a cover, such as a roof, in areas where contaminants are of concern so that rain or snow does not fall on the area and wash the contaminants downslope.

Inspect paving regularly for cracks that may allow contaminants to seep into the ground. Also, check to make sure that the drains receiving the storm water flow from the paved area remain unclogged with sediment or other debris so that low areas do not flood and wash over the areas where the contaminants may be.

The advantages and disadvantages of graded areas and pavement are presented below:

Advantages of Graded Areas and Pavement:

* Are effective in limiting storm water contact with contaminants.

* Are relatively inexpensive and easily implemented.

Disadvantages of Graded Areas and Pavement:

* May be uneconomical to regrade and resurface large areas.

* May not be effective during heavy precipitation.

8.3 EXPOSURE MINIMIZATION PRACTICES

By eliminating or minimizing the possibility of storm water coming into contact with pollutants, facilities can eliminate or minimize the contamination of storm water discharges associated with their industrial activity. As a result, fewer materials will be carried away by storm water runoff, the costs of collecting and treating contaminated storm water will be decreased, and safety and environmental liabilities that result from spills and leaks will be reduced.

Completely eliminating the exposure of materials to storm water is not always possible, however. For many industrial facilities, enclosure of facility grounds is not technologically or economically possible. Therefore, this section describes several simple and inexpensive structural and nonstructural BMPs that a facility can use to minimize the exposure of materials to storm water.

Containing spills is one of the primary methods of minimizing exposure of contaminants to storm water runoff. Spill containment is used for enclosing any drips, overflows, leaks, or other liquid material releases, as well as for isolating and keeping pollutant spills away from storm water runoff.

There are numerous spill containment methods, ranging from large structural barriers to simple, small drip pans. The benefits of each of these practices vary based on cost, need for maintenance, and size of the spill they are designed to control. This section describes several containment methods, including:

* Containment diking
* Curbing
* Drip pans
* Catch basins
* Sumps

Other practices commonly used to minimize exposure of contaminants are also discussed, including the following:

* Covering
* Vehicle positioning
* Loading and unloading by air pressure or vacuum

8.3.1 Containment Diking

Containment dikes are temporary or permanent earth or concrete berms or retaining walls that are designed to hold spills. Diking, one of the most common types of containment, is an effective method of pollution prevention for aboveground liquid storage tanks and rail car or tank truck loading and unloading areas. Diking can provide one of the best protective measures against the contamination of storm water because it surrounds the area of concern and holds the spill, keeping spill materials separated from the storm water outside of the diked area.

Diking can be used at any industrial facility, but is most commonly used for controlling large spills or releases from liquid storage areas and liquid transfer areas.

Containment dikes should be large enough to hold an amount equal to the largest single storage tank at the particular facility plus the volume of rainfall. For rail car and tank truck loading and unloading operations, the diked area should be capable of holding an amount equal to any single tank truck compartment.

Materials used to construct the dike should be strong enough to safely hold spilled materials. The materials used usually depend on what is available on site and the substance to be contained, and may consist of earth (i.e., soil or clay), concrete, synthetic materials (liners), metal, or other impervious materials. In general, strong acids and bases may react with metal containers, concrete, and some plastics, so where spills may consist of these substances, other alternatives should be considered. Some of the more reactive organic chemicals may also need to be contained with special liners. If there are any questions about storing chemicals in certain dikes because of their construction materials, refer to the Material Safety Data Sheets (MSDSs).

Containment dikes may need to be designed with impervious materials to prevent leaking or contamination of storm water, surface water, and groundwater supplies.

Similarly, uncontrolled overflows from diked areas containing spilled materials or contaminated storm water should be prevented to protect nearby surface waters or groundwaters. Therefore, dikes should have either pumping systems or vacuum trucks available to remove the spilled materials. When evaluating the performance of the containment system, you should pay special attention to the overflow system, since it is often the source of uncontrolled leaks. If overflow systems do not exist, accumulated storm water should be released periodically. Contaminated storm water should be treated prior to release. Mechanical parts, such as pumps or even manual systems (e.g., slide gates, stopcock valves), may require regular cleaning and maintenance.

When considering containment diking as a BMP, you should consult local authorities about any regulations governing construction of such structures to comply with local and state requirements. Facilities located in a floodplain should contact their local flood control authority to ensure that construction of the dikes is permitted.

Inspections of containment dikes should be conducted during or after significant storms or spills to check for washouts or overflows. In addition, regular checks of containment dikes (i.e., testing to ensure that dikes are capable of holding spills) is recommended. Soil dikes may need to be inspected on a more frequent basis.

Changes in vegetation, inability of the structure to retain storm water dike erosion, or soggy areas indicate problems with the dike's structure. Damaged areas should be patched and stabilized immediately, where necessary. Earthen dikes may require special maintenance of vegetation, such as mowing and irrigation.

The advantages and disadvantages of containment diking are presented below:

Advantages of Containment Diking:

* Contains spills, leaks, and other releases and prevents them from flowing into runoff conveyances, nearby streams, or underground water supplies.

* Permits materials collected in dikes to be recycled.

* Is a common industry practice for storage tanks and already required for certain chemicals.

Disadvantages of Containment Diking:

* May be too expensive for some smaller facilities.

* Requires maintenance.

* Could collect contaminated storm water, possibly resulting in infiltration of storm water to groundwater.

8.3.2 Curbing

Like containment diking, curbing is a barrier that surrounds an area of concern. Curbing functions in a similar way to prevent spills, leaks, etc., from being released to the environment by routing runoff to treatment or control areas. The terms curbing and diking are sometimes used interchangeably.

Because curbing is usually smallscale, it cannot contain large spills like diking can; however, curbing is common at many facilities in small areas where handling and transferring liquid materials occur.

Curbing can be used at all industrial facilities. It is particularly useful in areas where liquid materials are transferred and as a storm water runoff control.

As with diking, common materials for curbing include earth, concrete, synthetic materials, metal, or other impenetrable materials. Asphalt is also a common material used in curbing.

For maximum efficiency of curbing, spilled materials should be removed immediately to allow space for future spills. Curbs should have pumping systems, rather than drainage systems, for collecting spilled materials. Manual or mechanical methods, such as those provided by sump systems, can be used to remove the material. Curbing systems should be maintained through curb repair (patching and replacement).

When using curbing for runoff control, facilities should protect the berm by limiting traffic and installing reinforced berms in areas of concern.

Spills of materials that are stored within a curbed area can be tracked outside of that area when personnel and equipment leave the area. This tracking can be minimized by grading within the curbing to direct the spilled materials to a downslope side of the curbing. This will keep the materials away from personnel and equipment that pass through the area. It will also allow the materials to accumulate in one area, making cleanup much easier.

Inspections should also be conducted before forecasted rainfall events and immediately after storm events. If spilled or leaked materials are observed, cleanup should start immediately. This will prevent overflows and/or contamination of storm water runoff. In addition, prompt cleanup of materials will prevent dilution by rainwater, which can adversely affect recycling opportunities. Inspection of curbed areas should be conducted regularly, to clear clogging debris. Because curbing is sized to contain small spill volumes, maintenance should also be conducted frequently to prevent overflow of any spilled materials.

The advantages and disadvantages of curbing are presented below:

Advantages of Curbing:

* Is an excellent method to control runon.
* Is inexpensive.
* Is easily installed.
* Materials spilled within curbed areas can be recycled.
* Exists as a common industry practice.

Disadvantages of Curbing:

* Is not effective for holding large spills.
* May require more maintenance than diking.

8.3.3 Drip Pans

Drip pans are small depressions or pans used to contain very small volumes of leaks, drips, and spills that occur at a facility. Drip pans can be depressions in concrete, asphalt, or other impenetrable materials or they can be made of metals, plastic, or any material that does not react with the dripped chemicals. Drip pans can be temporary or permanent.

Drip pans are used to catch drips from valves, pipes, etc., so that the materials or chemicals can be cleaned up easily or recycled before they can contaminate storm water. Although leaks and drips should be repaired and eliminated as part of a preventive maintenance program, drip pans can provide a temporary solution where repair or replacement must be delayed. In addition, drip pans can be an added safeguard when they are positioned beneath areas where leaks and drips may occur.

Drip pans can be used at any industry where valves and piping are present and the potential for small-volume leakage and dripping exist.

When using drip pans, consider the location of the drip pan, weather conditions, the type of material to be used for the drip pan, and how it will be cleaned.

The location of the drip pan is important because drip pans must be inspected and cleaned frequently, and they must be easy to reach and remove. In addition, take special care to avoid placing drip pans in precarious positions such as next to walkways, on uneven pavement/ground, or sitting on pipelines. Drip pans in these locations are easily overturned and may present a safety hazard, as well as an environmental hazard.

Weather conditions are also important factors. Heavy winds and rainfall move or damage drip pans because of their small size and their light weight (if not built in). To prevent this, secure the pans by installing or anchoring them. Drip pans may be placed on platforms or behind wind blocks or tied down.

For drip pans to be effective, employees must pay attention to the pans and empty them when they are nearly full. Because of their small holding capacities, drip pans will easily overflow if not emptied. Also, recycling efforts can be affected if storm water accumulates in drip pans and dilutes the spilled material. It is important to have clearly specified and easily followed practices of reuse/recycle and/or disposal, especially the disposal of hazardous materials. Many facilities dump the drip pan contents into a nearby larger-volume storage container and periodically recycle the contents of the storage container.

In addition, frequent inspection of the drip pans is necessary due to the possibility of leaks in the pan itself or in piping or valves that may occur randomly or irregular slow drips that may increase in volume. Conduct inspections before forecasted rainfall events to remove accumulated materials and immediately after storm events to empty storm water accumulations.

The advantages and disadvantages of drip pans are presented below:

Advantages of Drip Pans:

* Are inexpensive.

* Are easily installed and simple to operate.

* Allow for reuse/recycle of collected material.

* Empty or discarded containers may be reused as drip pans.

Disadvantages of Drip Pans:

* Contain small volumes only.

* Must be inspected and cleaned frequently.

* Must be secured during poor weather conditions.

* Contents may be disposed of improperly unless facility personnel are trained in proper disposal methods.

8.3.4 Collection Basins

Collection basins, or storage basins, are permanent structures where large spills or contaminated storm water are contained and stored before cleanup or treatment. Collection basins are designed to receive spills, leaks, etc., that may occur and prevent these materials from being released to the environment. Unlike containment dikes, collection basins can receive and contain materials from many locations across a facility.

Collection basins are commonly confused with treatment units such as ponds, lagoons, and other containment structures. Collection basins differ from these structures because they are designed to temporarily store storm water rather than treat it.

Collection basins are appropriate for all industrial sites where space allows. Collection basins are particularly useful for areas that have a high spill potential.

The design and installation considerations for collection basins include sizing the basin either to hold a certain amount of spill or a certain size storm, or both. In designing the collection system, the type of material for the conveyances, compatibility of various materials to be carried through the system, and requirements for compliance with State and local regulations should be considered. Ideally, the system should function to route the materials quickly and easily to the collection basin.

When spills occur, the collection system must route the spill or storm water immediately to the collection basin. After a spill is contained, the collection system and basin may require cleaning. Remove the collection basin contents immediately to prevent an unintentional release and recycle the spilled material as much as possible. Inspect the structure on a regular basis and after storm events or spills. Depending upon the types of pollutants that may be in the storm water, or are collected as spills, design of the basin may require a liner to prevent infiltration into the groundwater. Make sure that the installation of this BMP does not violate State groundwater regulations.

If it is possible that the collection basin may handle combustible or flammable spilled materials, explosion-proof pumping equipment and controls or other appropriate precautions should be taken to prevent explosions or fires. Consult OSHA and local safety codes and standards for specific requirements and guidance.

The advantages and disadvantages of collection basins are presented below:

Advantages of Collection Basins:

* Can store contaminated storm water until directed to a treatment facility.

* Can collect spills for recycling where materials are separated.

Disadvantages of Collection Basins:

* May need a conveyance system for increased effectiveness.

* May collect materials that are not compatible.

* May reduce the potential for recycling materials by collecting storm water, which dilutes the materials.

* May create groundwater problems if pollutants infiltrate into the ground.

8.3.5 Sumps

Sumps are holes or low areas that are structured so that liquid spills or leaks will flow down toward a particular part of a containment area. Frequently, pumps are placed in a depressed area and are turned on automatically to transfer liquids away from the sump when the level of liquids gets too high. Sumps can be temporary or permanent.

Sumps can be used at all facilities. Sumps are used with other spill containment and treatment measures and can be located almost anywhere on site. Sumps are frequently located in low-lying areas within material-handling or storage areas.

When designing and installing a sump system, consider the pump location, function, and system alarms. Design and install the sump in the lowest-lying area of a containment structure, allowing for materials to gather in the area of the sump. Construct the sump of impenetrable materials and provide a smooth surface so that liquids are funneled toward the pump. It may be appropriate to house the pumps in a shed or other structure for protection and stabilization.

There are numerous factors that should be considered when purchasing a pump. Base the size of the pump on the maximum expected volume to be collected in the containment structure. In some cases, more than one pump may be appropriate. Typically, pumps that can be submerged under the spill are the most appropriate for areas where large spills may occur and that may submerge the sump area. The viscosity (thickness) of the material and the distance that the material must be pumped are also important considerations. Install pumps according to the manufacturer's recommendations.

An alarm system can be installed for pumps that are used to remove collected materials. An alarm system can indicate that a pump should be operated by hand or that an automatically operated pump has failed to function. Ultimately, facility personnel should have some mechanism to take action to prevent spills from bypassing and overflowing containment structures.

The pumps and the alarm system used in the sump generally require regular inspections for service and maintenance of parts based on manufacturers' recommendations.

If it is possible that the sump may handle combustible or flammable spilled materials, explosionproof pumping equipment and controls or other appropriate precautions should be taken to prevent explosions or fires. Consult OSHA and local safety codes and standards for specific requirements and guidance.

The advantages and disadvantages of sumps are presented below:

Advantages of Sumps:

* Provide a simple and quick collection method for recycling, reusing, or treating materials in a containment structure.

* Are commonly used at industrial facilities.

Disadvantages of Sumps:

* Pumps may clog easily if not designed correctly.

* May require maintenance/servicing agreements with pump dealers.

* Costs for purchasing and/or replacing pumps may be high.

8.3.6 Covering

Covering is the partial or total physical enclosure of materials, equipment, process operations, or activities. Covering certain areas or activities prevents storm water from coming into contact with potential pollutants and reduces material loss from wind blowing. Tarpaulins, plastic sheeting, roofs, buildings, and other enclosures are examples of covering that are effective in preventing storm water contamination. Covering can be temporary or permanent.

Covering is appropriate for outdoor material storage piles (e.g., stockpiles of dry materials, gravel, sand, compost, sawdust, wood chips, deicing salt, and building materials) and areas where liquids and solids in containers are stored or transferred. Although it may be too expensive to cover or enclose all industrial activities, cover high-risk areas (identified during the storm water pollutant source identification). For example, cover chemical preparation areas, vehicle maintenance areas, areas where chemically treated products are stored, and areas where salts are stored.

If covering or enclosing the entire activity is not possible, the high-risk part of the activity can often be separated from other processes and covered. Another option that reduces the cost of building a complete enclosure is to build a roof over the activity. A roof may also eliminate the need for ventilation and lighting systems.

Evaluate the strength and longevity of the covering, as well as its compatibility with the material or activity being enclosed. When designing an enclosure, consider access to materials, their handling, and transfer. Materials that pose environmental and safety dangers because they are radioactive, biological, flammable, explosive, or reactive require special ventilation and temperature considerations.

Covering alone may not protect exposed materials from storm water contact. Place the material on an elevated, impermeable surface or build curbing around the outside of the materials to prevent problems from runon of uncontaminated storm water from adjacent areas.

Frequently inspect covering, such as tarpaulins, for rips, holes, and general wear. Anchor the covering with stakes, tie-down ropes, large rocks, tires, or other easily available heavy objects.

Practicing proper materials management within an enclosure or underneath a covered area is essential. For example, floor drainage within an enclosure should be properly designed and connected to the wastewater sewer where appropriate and allowed. If connection to an off-site wastewater sewer is considered, the local publicly owned treatment works (POTW) should be consulted to find out if there are any pretreatment requirements or restrictions that must be followed.

The advantages and disadvantages of covering are presented below:

Advantages of Covering:

* Is simple and effective.

* Is commonly inexpensive.

Disadvantages of Covering:

* Requires frequent inspection.

* May pose health or safety problems if enclosure is built over certain activities.

8.3.7 Vehicle Positioning

Vehicle positioning is the practice of locating trucks or rail cars while transferring materials to prevent spills of materials onto the ground surface, which may then contaminate storm water runoff. Vehicle positioning is a simple and effective method of material spill prevention and yet it is commonly overlooked.

Vehicle positioning can be used at all types of industrial facilities. This practice is appropriate for any area where materials are transferred from or to vehicles, such as loading and unloading areas, storage areas, and material transfer areas. Use vehicle positioning in conjunction with other practices such as covering, sumps, drip pans, or loading and unloading by air pressure or vacuum where chemical spills are of concern.

The purpose of vehicle positioning is to locate vehicles in a stable and appropriate position to prevent problems, such as spills resulting from broken material storage containers, spills caused by vehicle movement during materials transfer activities, and spills caused by improperly located vehicles. Vehicles should also be positioned near containment or flow diversion systems to collect unexpected spills from leaks in transfer lines or connections. The following activities are included in this practice:

* Constructing walls that help in positioning the vehicles.

* Positioning vehicle either over a drain or on a sloped surface that drains to a containment structure.

* Outlining required vehicle positions on the pavement.

* Using wheel guards or wheel blocks.

* Posting signs requiring the use of emergency brakes.

* Requiring vehicles to shut off engines during materials transfer activities.

The advantages and disadvantages of vehicle positioning are presented below:

Advantages of Vehicle Positioning:

* Is inexpensive.

* Is easy and effective.

Disadvantages of Vehicle Positioning:

* May require redesign of loading and unloading areas.

8.3.8 Loading and Unloading by Air Pressure or Vacuum

Air pressure and vacuum systems are commonly used for transporting and loading and unloading materials. These systems are simple to use and effective in transferring dry chemicals or solids from one area to another, but are less effective as the particles of material become more dense.

In an air pressure system, a safety relief valve and a dust collector are used to separate the dry materials from the air and then release the air accumulated during transfer operations. In a vacuum system, a dust collection device and an air lock, such as a rotary gate or trap door feeder, are typically used.

The use of mechanical equipment that involves enclosed lines, such as those provided by air pressure (also referred to as pneumatic) and vacuum loading systems, are effective methods for minimizing releases of pollutants into the environment. Because of the enclosed nature of the system, pollutants are not exposed to wind or precipitation and therefore have less potential to contaminate storm water discharges.

Air pressure and vacuum systems can be used at all types of industrial facilities. This equipment is located in material-handling areas to use for storing, loading and unloading, transporting, or conveying materials.

Unlike many of the other BMPs discussed in this book, air pressure and vacuum systems may be expensive because of the costs of purchasing the system and retrofitting the system to existing materials-handling procedures. In many cases, these systems can be shipped to a facility and be installed on site without contractor help. Manufacturers' recommendations should be followed closely to ensure proper installation. In other cases, systems may have to be designed specifically for a site. Proper design and installation are very important for air pressure and vacuum systems to be as effective as possible. The equipment may be weatherproof or, if not, consider enclosing or covering the equipment.

Conduct routine inspections of air pressure and vacuum systems. Regular maintenance is required of these systems, especially the dust collectors. Conduct maintenance activities based on manufacturers' recommendations. Inspect air pressure systems more frequently due to the greater potential for leaks to the environment.

The advantages and disadvantages of loading and unloading by air pressure or vacuum are presented below:

Advantages of Loading and Unloading by Air Pressure or Vacuum:

* Are quick and simple.

* May be economical if materials can be recovered.

* Will minimize exposure of pollutants to storm water.

Disadvantages of Loading and Unloading by Air Pressure or Vacuum:

* May be costly to install and maintain.

* May not be appropriate for some denser materials.

* May require site-specific design.

* Dust collectors may need a Clean Air Act permit for installation.

8.4 MITIGATION PRACTICES

Mitigation involves cleaning up or recovering a substance after it has been released or spilled to reduce the potential impact of a spill before it reaches the environment. Therefore, pollution mitigation is a second line of defense where pollution prevention practices have failed or are impractical. Because spills cannot always be avoided at industrial sites, it is necessary to plan for these events and to design proper response procedures. This section discusses mitigation BMPs to avoid contamination of storm water. Most of the mitigation practices discussed are simple and should be incorporated in good housekeeping and spill response plans for the facility. The mitigation practices discussed include manual cleanup methods, such as sweeping and shoveling, mechanical cleanup by excavation or vacuuming, and cleanup with sorbents and gels.

Facilities are cautioned that spills of certain toxic and hazardous substances and their cleanup may be covered under regulations, including those imposed under the Superfund Amendments and Reauthorization Act, the Comprehensive Environmental Response, Compensation, and Liability Act, and the Resource Conservation and Recovery Act.

8.4.1 Sweeping

Sweeping with brooms, squeegees, or other mechanical devices is used to remove small quantities of dry chemicals and dry solids from areas that are exposed to precipitation or storm water runoff. These areas may include dust- or contaminant-covered bags, drums containing remaining materials on their lids, areas housing enclosed or covered materials, and spills of dry chemicals and dry solids in locations on the industrial site. Cleaning by sweeping with brooms is a low-cost practice that can be performed by all employees and requires no special equipment or training.

Sweeping can be used at many material-handling areas and process areas in all types of industrial facilities. Timing is an important consideration for all mitigation practices. To be effective as a storm water control, cleanup must take place before rainfall or contact with storm water runoff or before an outside area is hosed down.

Do not limit cleanup activities to those outside activities that are exposed to rainfall. In many cases, tracking of materials to the outside from areas that are enclosed or covered (e.g., on shoes) may also occur.

Store brooms appropriately and do not expose them to precipitation. In addition, rules of compatibility also apply. Do not use the same broom to clean up two chemicals that are incompatible. Determine the compatibility between the brooms themselves and the chemical of concern before using this practice. In some instances, chemicals should be vacuumed instead of swept. Be sure that swept material is disposed of properly.

The advantages and disadvantages of sweeping are presented below:

Advantages of Sweeping:

* Is inexpensive.
* Requires no special training.
* Provides recycling opportunities.

Disadvantages of Sweeping:

* Is a labor-intensive practice.
* Is limited to small releases of dry materials.

8.4.2 Shoveling

Shoveling is another manual cleanup method that is simple and low in cost. Generally, shoveling can be used to remove larger quantities of dry chemicals and dry solids, as well as to remove wetter solids and sludge. Shoveling is also useful in removing accumulated materials from sites not accessible by mechanical cleanup methods.

Shoveling can be used at any facility. Shoveling provides an added advantage over sweeping because cleanup methods are not limited to dry materials. In many cases, accumulated solids and sludges that are in ditches, sumps, or other facility locations can be effectively and quickly removed by shoveling.

Shovels can also be used to clean up contaminated snows. Timing is an important consideration in any mitigation practice. Materials that could contaminate storm water runoff should be removed before any storm event.

As with brooms, clean and store shovels properly. Also, consider planning for the transport and disposal or reuse of the shoveled materials.

The advantages and disadvantages of shoveling are presented below:

Advantages of Shoveling:

* Is inexpensive.
* Provides recycling opportunities.
* Can remediate larger releases and is effective for dry and wet materials.

Disadvantages of Shoveling:

* Is labor intensive.
* Is not an appropriate practice for large spills.

8.4.3 Excavation Practices

Excavation (i.e., removal of contaminated material) of released materials is typically conducted by mechanical equipment, such as plows and backhoes. Generally, plowing and backhoeing can be done using a specifically designed vehicle, tractor, or truck.

Excavation removes the materials of concern and any deposition of contaminants, thereby reducing the potential for storm water contamination. Mechanical cleanup methods are typically less precise than manual cleanup methods, resulting in reduced opportunities for recycle and reuse.

Excavation practices are most useful for large releases of dry materials and for areas contaminated by liquid material releases. In excavation, you want to be sure that all of the contaminated material is removed.

Timing is an important consideration for all mitigation practices. To be effective as a storm water control, cleanup must take place before a rainfall event.

Conduct inspections and operations and maintenance in accordance with a manufacturer's recommendations, which may include the following:

* A specified frequency for inspection, maintenance, and servicing of the equipment.

* Parts replacement, rotation, and lubrication specifications.

* Procedures for evaluating all parts.

As with any equipment used during cleanup, other considerations apply, including the following:

* Plows, backhoes, etc., should be stored appropriately with no exposure to precipitation.

* Excavated materials should be properly handled or disposed of.

The advantages and disadvantages of excavation practices are presented below:

Advantages of Excavation Practices:

* Are a cost-effective method for cleaning up dry materials released.

* Are common and simple.

Disadvantages of Excavation Practices:

* Are less precise, resulting in less recycling and reuse opportunities.

8.4.4 Vacuum and Pump Systems

Vacuum and pump systems are effective for cleaning up spilled or exposed materials. The benefits of vacuum and pump cleanup systems include simplicity and speed. With such systems, only the spilled materials need be collected. Also, these systems are often portable and can be used at many locations to clean up releases to the environment. Portable systems can usually be rented.

Vacuum and pump systems can be used at any industrial facility. Both wet and dry materials can be collected with these systems. Vacuum systems can be used in material-handling areas and process areas.

Consider the area of use and the most appropriate size for the system. Since these systems can be portable, size is important, especially if materials will be stored in the unit. In this case, the portable system must have enough suction or positive air pressure to transport materials over long distances. Include plans for proper disposal or reuse of the collected materials.

The advantages and disadvantages of vacuum and pump systems are presented below:

Advantages of Vacuum and Pump Systems:

* Remove materials by air pressure or vacuum quickly and simply.

* Collect materials accurately.

* Offer good recycling opportunities.

Disadvantages of Vacuum and Pump Systems:

* May require high initial capital cost.

* Require equipment maintenance.

8.4.5 Sorbents

Sorbents are materials that are capable of cleaning up spills through the chemical processes of adsorption and absorption. Sorbents adsorb (an attraction to the outer surface of a material) or absorb (taken in by the material like a sponge) only when they come in contact with the spill materials. The sorbents must be mixed with a spill or the liquid must be passed through the sorbent. Sorbent materials come in many different forms from particles to foams. Often the particles are held together in structures called booms, pads, or socks. Sorbents include, but are not limited to, the following:

* *Common Materials (clays, sawdust, straw, and flyash)*: Generally come in small particles that can be thrown onto a spill that is on a surface. The materials absorb the spill by taking up the liquid.

* *Polymers (polyurethane and polyolefin):* Come in the form of spheres, beads, or foam tablets. These materials absorb a chemical spill by taking up the liquid into their open-pore structure.

* *Activated Carbon:* Comes in a powdered or granular form and can be mixed with liquids to remove pollutants. This sorbent works by adsorbing the organics to its surface and can be recycled and then reused by a process called regeneration.

* *"Universal Sorbent Material":* Is a silicate glass foam consisting of rounded particles that can absorb the material.

Sorbents are useful BMPs for facilities with liquid materials on site. Timing is important for these practices. To be effective as a storm water BMP, cleanup must take place before a rainfall. Sorbents are often used in conjunction with curbing to provide cleanup of small spills within a containment area.

"Universal Sorbent Materials" are suitable for use on many compounds including acids, alkalis, alcohols, aldehydes, arsenate, ketones, petroleum products, and chlorinated solvents.

Activated carbon is useful for adsorbing many organic compounds. Organics that are diluted in water can be passed through a column that is filled with the activated carbon material to remove the organics, or the activated carbon can be mixed into the water and can then be filtered out.

Polyurethane is good with chemical liquids such as benzene, chlorinated solvents, epichlorohydrin, and phenol. Polyolefin is used to remove organic solvents, such as phenol and various chlorinated solvents. The beads and spheres are usually mixed into a spill by use of a blower and then are skimmed from the top surface by use of an oil boom.

More common materials such as clay, sawdust, straw, and fly-ash can be used for a liquid spill on a surface that is relatively impenetrable, and are usually spread over the spill area with shovels.

Booms, pads, and socks are also useful in areas where there are small liquid spills or drips or where small amounts of solids may mix with small amounts of storm water runoff. They can function both to absorb the pollutants from the storm water and restrict the movement of a spill. Socks are often used together with curbing to clean up small spills.

Because sorbents work by a chemical or physical reaction, some sorbents are better than others for certain types of spills. Therefore, the use of sorbents requires that personnel know the properties of the spilled material(s) to know which sorbent is appropriate. To be effective, sorbents must adsorb the material spilled, but must not react with the spilled material to form hazardous or toxic substances. Follow the manufacturers' recommendations.

For sorbents to be effective, they must be applied immediately in the release area. The use of sorbent material is generally very simple: the sorbent is added to the area of release, mixed well, and allowed to adsorb or absorb. Many sorbents are not reusable once they have been used. Proper disposal is required.

The advantages and disadvantages of sorbents are presented below:

Advantages of Sorbents:

* Work in water environments (booms and socks).
* Offer recycling opportunities (some types of sorbents).

Disadvantages of Sorbents:

* Require a knowledge of the chemical makeup of a spill (to choose the best sorbent).
* Offer no recycling opportunities (some types of sorbents).
* May be expensive practice for large spills.
* May create disposal problems and increase disposal costs by creating a solid waste and potentially a hazardous waste.

8.4.6 Gelling Agents

Gelling agents are materials that interact with liquids either physically or chemically (i.e., thickening or polymerization). Some of the typical gelling agents are polyelectrolytes, polyacrylamide, butylstyrene copolymers, polyacrylonitrile, polyethylene oxide, and a gelling agent referred to as the universal gelling agent, which is a combination of these synthetics.

Gelling interacts with a material by concentrating and congealing it to become semisolid. The semisolid gel later forms a solid material, which can then be cleaned up by manual or mechanical methods. The BMP of using a gelling agent is one of the few ways to effectively control a liquid spill before it reaches a receiving water or infiltrates into the soil and then groundwater.

Gelling agents are useful for facilities with significant amounts of liquid materials stored onsite. **Gels cannot be used to**

clean up spills on surface water unless authorized by the U.S. Coast Guard or EPA Regional Response Team.

Gels can be used to stop the liquid's flow on land, prevent its seeping into the soil, and reduce the surface spreading of a spill. Because of these properties, gels can reduce the need for extensive cleanup methods and reduce the possibility of storm water contamination from an uncontrolled industrial spill. As with sorbents, the use of gels simply involves the addition of the gel to the area of the spill, mixing well, and allowing the mass to congeal. To use gels correctly, however, personnel need to know the properties of the spilled materials so that they can choose the correct gel.

Timing is particularly important for gelling agent use. To prevent the movement of materials, gelling agents must be applied immediately after the spill. The use of gelling agents results in a large bulk of congealed mass that usually cannot be separated. Ultimately, this mass will need to be cleaned up by manual or mechanical methods and disposed of properly.

The advantages and disadvantages of gelling agents are presented below:

Advantages of Gelling Agents:

* Stop the movement of spilled or released liquid materials.

* Require no permanent structure.

Disadvantages of Gelling Agents:

* May require knowledge of the spilled materials to select correct gelling agents.

* Usually offer no recycling opportunities.

* May be difficult to clean up.

* May create disposal problems and increase disposal costs by creating a solid waste and potentially a hazardous waste.

8.5 ADDITIONAL PREVENTIVE PRACTICES

A number of other preventive measures can be taken at industrial sites to limit or prevent the exposure of storm water runoff to contaminants. This section describes a few of the most easily implemented measures:

* Preventive monitoring practices
* Dust control (land disturbance and demolition areas)
* Dust control (industrial)
* Signs and labels
* Security
* Area control procedures
* Vehicle washing

8.5.1 Preventive Monitoring Practices

Preventive monitoring practices include the routine observation of a process or piece of equipment to ensure its safe performance. It may also include the chemical analysis of storm water before discharge to the environment.

Automatic Monitoring System: In areas where overflows, spills, and catastrophic leaks are possible, an automatic monitoring system is recommended. Some Federal, State, and local laws require such systems to be present if threats exist to the health and safety of personnel and the environment. For material management areas, monitoring may include liquid level detectors, pressure and temperature gauges, and pressure-relief devices. In material transfer, process, and material-handling areas, automatic monitoring systems can include pressure drop shutoff devices, flow meters, thermal probes, valve position indicators, and operation lights. Loading and unloading operations might use these devices for measuring the volume of tanks before loading, for weighing vehicles or containers, and for determining rates of flow during loading and unloading.

Automatic Chemical Monitoring: Measures the quality of plant runoff to determine whether discharge is appropriate or if diversion to a treatment system is warranted. Such systems might monitor pH, turbidity, or conductivity. These parameters might be monitored in diked areas, sewers, drainage ditches, or holding ponds. Systems can also be designed to signal automatic diversion of contaminated storm water runoff to a holding pond (e.g., a valve or a gate could be triggered by a certain pollutant in the storm water runoff).

Manned Operations: In material-transfer areas and process areas, personnel can be stationed to watch over the operations so that any spills or mismanagement of materials can be corrected immediately. This is particularly useful at loading and unloading areas where vehicles or equipment must be maneuvered into the proper position to unload.

Nondestructive Testing: Some situations require that storage tank or a pipeline system be tested without being physically moved or disassembled. The structural integrity of tanks, valves, pipes, joints, welds, and other equipment can be tested using nondestructive methods. Acoustic emission tests use high-frequency sound waves to draw a picture of the structure to reveal cracks, malformations, or other structural damage. Another type of testing is hydrostatic pressure testing. During pressure testing, the tank or pipe is subjected to pressures several times the normal pressure. A loss in pressure during the testing may indicate a leak or some other structural damage. Tanks and containers should be pressure tested as required by federal, state, or local regulations.

Automated monitoring systems should be placed in an area where plant personnel can easily observe the measurements. Alarms can be used in conjunction with the measurement display to warn personnel. Manned operations should have communication systems available for getting help in case spills or leaks occur. Especially sensitive or spill-prone areas may require backup instrumentation in case the primary instruments malfunction.

Mechanical and electronic equipment should be operated and maintained according to the manufacturers' recommendations. Equipment should be inspected regularly to ensure proper and accurate operation.

The pollution prevention team, in consultation with a certified safety inspector, should evaluate system monitoring requirements to decide which systems are appropriate based on hazard potential.

Advantages of Preventive Monitoring Practices:

* Pressure and vacuum testing can locate potential leaks or damage to vessels early. The primary benefit of such testing is in ensuring the safety of personnel, but it also has secondary benefits including prevention of storm water contamination.

* Automatic system monitors allow for early warnings if a leak, overflow, or catastrophic incident is imminent.

* Manning operations, especially during loading and unloading activities, is effective and generally inexpensive.

* The primary benefit of nondestructive testing is in ensuring the safety of personnel, but it also has secondary benefits including early detection of the potential for contaminating storm water runoff.

Disadvantages of Preventive Monitoring Practices:

* Plant personnel often do not have the expertise to maintain automatic equipment.

* Automatic equipment can fail without warning.

* Automated process control and monitoring equipment may be expensive to purchase and operate.

8.5.2 Dust Control (Industrial)

Dust controls for material-handling areas are controls that prevent pollutants from entering storm water discharges by reducing the surface and air transport of dust caused by industrial activities. Consider the following types of controls:

* Water spraying
* Negative pressure systems (vacuum systems)
* Collector systems (bag and cyclone)
* Filter systems
* Street sweeping

The purpose of industrial dust control is to collect or contain dusts to prevent storm water runoff from carrying the dusts to the sewer collection system or to surface waters.

Dust control is useful in any process area, loading and unloading area, material-handling areas, and transfer areas where dust is generated. Street sweeping is limited to areas that are paved.

Mechanical dust-collection systems are designed according to the size of dust particles and the amount of air to be processed. Manufacturers' recommendations should be followed for installation, as well as the design of the equipment.

If water sprayers are used, dust-contaminated waters should be collected and taken for treatment. Areas will probably need to be resprayed to keep dust from spreading.

Two kinds of street sweepers are common: brush and vacuum. Vacuum sweepers are more efficient and work best when the area is dry.

Mechanical equipment should be operated according to the manufacturers' recommendations and should be inspected regularly.

Advantages of Dust Control (Industrial):

* May cause a decrease of respiratory problems in employees around the site.

* May cause less material to be lost and may, therefore, save money.

* Provides efficient collection of larger dust particles (street sweepers).

Disadvantages of Dust Control (Industrial):

* Is generally more expensive than manual systems.

* May be impossible to maintain by plant personnel (the more elaborate equipment).

* Is labor and equipment intensive and may not be effective for all pollutants (street sweepers).

8.5.3 Dust Control at Land-Disturbance and Demolition Areas

Dust controls for land-disturbance and demolition areas are any controls that reduce the potential for particles being carried through air or water. Types of dust control are:

* *Irrigation:* Irrigation is a temporary measure involving a light application of water to moisten the soil surface. The process should be repeated as necessary.

* *Minimization of Denuded Areas:* Minimizing soil exposure reduces the amount of soil available for transport and erosion. Soil exposure can be lessened by temporary or permanent soil stabilization controls, such as seeding, mulching, topsoiling, crushed stone or coarse gravel spreading, or tree planting. Maintaining existing vegetation on a site will also help control dust.

* *Wind Breaks:* Wind breaks are temporary or permanent barriers that reduce airborne particles by slowing wind velocities (slower winds do not suspend particles). Leaving existing trees and large shrubs in place will create effective wind breaks. More temporary types of wind breaks are solid board fences, snow fences, tarp curtains, bales of hay, crate walls, and sediment walls.

* *Tillage:* Deep plowing will roughen the soil surface to bring up to the surface cohesive clods of soil, which in turn rest on top of dusts, protecting them from wind and water erosion. This practice is commonly practiced in arid regions where establishing vegetation may take time.

* *Chemical Soil Treatments (palliatives):* These are temporary controls that are applied to soil surfaces in the form of spray-on adhesives, such as anionic asphalt emulsion, latex emulsion, resin-water emulsions, or calcium chloride. The palliative is the chemical used. These should be used with caution as they may create pollution if not used correctly.

Dust controls can be used on any site where dust may be generated and where the dust may cause on-site and off-site damage. Dust controls are especially critical in arid areas, where reduced rainfall levels expose soil particles for transport by air and runoff. This control should be used in conjunction with other sedimentation controls such as sediment traps.

To control dust during land disturbance and at demolition areas, exposure of soil should be limited as much as possible. When possible, work that causes soil disturbance or involves demolition should be done in phases and should be accompanied by temporary stabilization measures. These precautions will minimize the amount of soil that is disturbed at any one time and, therefore, control dust.

Oil should not be used to control dust because of its high potential for polluting storm water discharges.

Irrigation will be most effective if site drainage systems are checked to ensure that the right amount of water is used. Too much water can cause runoff problems.

Chemical treatment is only effective on mineral soils, as opposed to muck soils, because the chemicals bond better to mineral soils. Therefore, it should be used only in arid regions. Vehicular traffic should be routed around chemically treated areas to avoid tracking of the chemicals. Certain chemicals may be inappropriate for some types of soils or application areas. For example, spraying chemicals on the soil of an industrial site near a school may be dangerous. Local governments usually have information about restrictions on the types of palliatives that may be used. Special consideration must be given to preserving groundwater quality whenever chemicals are applied to the land.

Since most of these techniques are temporary controls, sites should be inspected often and materials should be reapplied when needed. The frequency for these inspections depends on sitespecific conditions, weather conditions, and the type of technique used.

Advantages of Dust Control at Land-Disturbance and Demolition Areas:

* Can help prevent wind- and water-based erosion of disturbed areas and will reduce respiratory problems in employees.

* Some types can be implemented quickly at low cost and effort (except wind breaks).

* Helps preserve the aesthetics of the site and screens certain activities from view (wind breaks).

* Vegetative wind breaks are permanent and an excellent alternative to chemical use.

Disadvantages of Dust Control at Land-Disturbance and Demolition Areas:

* Some types are temporary and must be reapplied or replenished regularly.

* Some types are expensive (irrigation and chemical treatment) and may be ineffective under certain conditions.

* May result in health and/or environmental hazards, e.g., if overapplication of the chemicals leaves large amounts exposed to wind and rain erosion or groundwater contamination.

* May create excess runoff that the site was not designed to control (irrigation).

* May cause increased off-site tracking of mud (irrigation).

* Is not as effective as chemical treatment or mulching and seeding; requires land space that may not be available at all locations (wind breaks).

8.5.4 Signs and Labels

Signs and labels identify problem areas or hazardous materials at a facility. Warning signs, often found at industrial facilities, are a good way to suggest caution in certain areas. Signs and labels can also provide instructions on the use of materials and equipment. Labeling is a good way to organize large amounts of materials, pipes, and equipment, particularly on large sites.

Labels tell material type and container contents. Accurate labeling can help facilities to quickly identify the type of material released so facility personnel can respond correctly.

Two effective labeling methods include color coding and Department of Transportation (DOT) labeling. Color coding is easily recognized by facility personnel and simply involves painting/coating or applying an adhesive label to the container. Color codes must be consistent throughout the facility to be effective, and signs explaining the color codes should be posted in all areas.

DOT requires that labels be prominently displayed on transported hazardous and toxic materials. Labeling required by

DOT could be expanded to piping and containers, making it easy to recognize materials that are corrosive, radioactive, reactive, flammable, explosive, or poisonous.

Signs and labels can be used at all types of facilities. Areas where they are particularly useful are material transfer areas, equipment areas, loading and unloading areas, or anywhere information might prevent contaminants from being released to storm water.

Signs and labels should be visible and easy to read. Useful signs and labels might provide the following information:

* Names of facility and regulatory personnel, including emergency phone numbers, to contact in case of an accidental discharge, spill, or other emergency.

* Proper uses of equipment that could cause release of storm water contaminants.

* Types of chemicals used in high-risk areas.

* The direction of drainage lines/ditches and their destination (treatment or discharge).

* Information on a specific material.

* Refer to OSHA standards for sizes and numbers of signs required for hazardous material labeling.

Hazardous chemicals might be labeled as follows:

* Danger
* Combustible
* Warning
* Caution
* Flammable
* Poisonous
* Caustic
* Corrosive
* Volatile
* Explosive

Periodic checks can ensure that signs are still in place and labels are properly attached. Signs and labels should be replaced and repaired as often as necessary.

Advantages of Signs and Labels:

* Are inexpensive and easily used.

Disadvantages of Signs and Labels:

* Must be updated and maintained so they are legible.

8.5.5 Security

Setting up a security system as part of your plan could help prevent an accidental or intentional release of materials to storm water runoff as a result of vandalism, theft, sabotage, or other improper uses of facility property. If your facility already has a security system, consider improving it by training security personnel about the specifics of the Storm Water Pollution Prevention Plan. Routine patrol, lighting, and access control are discussed below as possible measures to include in your facility's security system. Routine patrol, lighting, and access control are measures that can be used at any facility.

Security information could be included in the existing training required by the plan to instruct personnel about where and how to patrol areas within the facility. Instruction might also include what to look for in problem areas and how to respond to problems. During routine patrol, security personnel can actively search the facility site for indications of spills, leaks, or other discharges; respond to any disturbance resulting from intruders or inappropriate facility operations; and generally work as a safeguard to prevent unexpected events. Routine patrols could be an effective part of the Storm Water Pollution Prevention Plan, especially for large facilities with established security measures. To make this practice effective, security personnel can help develop the Storm Water Pollution Prevention Plan, possibly with one person acting as a member of the pollution prevention committee.

Sufficient lighting throughout the facility during daytime and night hours will make it easier to get to equipment during checks and will make it easy to detect spills and leaks that might otherwise be hidden. Routine patrols are also easier with proper lighting.

Controlling access to the industrial site is an important part of plant security and of activity and traffic control. Signs, fencing, guard houses, dog patrols, and visitor clearance requirements are often used to control site access.

* *Signs* are the simplest, most inexpensive method of access control, but they are limited in their actual control since they provide no physical barriers and require that people obey them voluntarily.

* *Fencing* provides a physical barrier to the facility site and an added means of security.

* *Guard houses* used with visitor rules can help to ensure that only authorized personnel enter the facility site and can limit vehicular traffic as well.

* *Traffic signs* are also useful at facility sites. Restricting vehicles to paved roads and providing direction and warning signs can help prevent accidents. Where restricting vehicles to certain pathways is not possible, it is important to ensure that all aboveground valves and pipelines are well marked.

Advantages of Security:

* Provides a preventive safeguard to operational malfunctions or other facility disturbances (routine patrols).

* Allows easier detection of vandals or thieves (lighting).

* Allows easier detection of spills, leaks, or other releases (lighting).

* Prevents spills by providing good visibility (lighting).

* Prevents unauthorized access to facility (access control).

Disadvantages of Security:

* May not be feasible for smaller facilities.

* May be costly (e.g., installation of lighting systems).

* May increase energy costs as a result of additional lighting.

* May not be feasible to have extensive access controls at smaller facilities.

8.5.6 Area Control Procedures

The activities conducted at an industrial site often result in the materials being deposited on clothes and footwear and then being carried throughout the facility site. As a result, these materials may find their way into the storm water runoff.

Area control procedures involve practicing good housekeeping measures such as maintaining indoor or covered material storage and industrial processing areas. If the area is kept clean, the risk of accumulating materials on footwear and clothing is reduced. In turn, the chance of leftover pollutants making contact with storm water and polluting surface water is minimized.

Area control measures can be used at any facility where materials may be tracked into areas where they can come in contact with storm water runoff. Areas can include material-handling areas, storage areas, or process areas.

Materials storage areas and industrial processing areas should be checked regularly to ensure that good housekeeping measures are being implemented. Cover-garments, foot mats, and other devices used to collect residual material near the area should be cleaned regularly.

Other effective practices include the following:

* Brushing off clothing before leaving the area.

* Stomping feet to remove material before leaving the area.

* Using floor mats at area exits.

* Using coveralls, smocks, and other overgarments in areas where exposure to material is of greatest concern (employees should remove the overgarments before leaving the area).

* Posting signs to remind employees about these practices.

Advantages of Area Control Procedures:

* Are easy to implement.

* Result in a cleaner facility and improved work environment.

Disadvantage of Area Control Procedures:

* May be seen as tedious by employees and therefore may not be followed.

Chapter 9

EROSION PREVENTION, SEDIMENT CONTROL, AND INFILTRATION PRACTICES

9.1 INTRODUCTION

This chapter describes additional site-specific Best Management Practices (BMPs) designed to minimize, reduce, and eliminate storm water contamination from various industrial activities. Several different erosion prevention, sediment control, and infiltration practices are provided to help in reducing storm water contamination at industrial facility sites.

9.2 VEGETATIVE EROSION PREVENTION AND SEDIMENT CONTROL PRACTICES

Any site where soils are exposed to water, wind, or ice can have soil erosion and sedimentation problems. Erosion is a natural process in which soil and rock material is loosened and removed. Sedimentation occurs when soil particles are suspended in surface runoff or wind and are deposited in streams and other water bodies.

Human activities can accelerate erosion by removing vegetation, compacting or disturbing the soil, changing natural drainage patterns, and by covering the ground with impermeable surfaces (pavement, concrete, buildings). When the land surface is developed or "hardened" in this manner, storm water and snowmelt can not seep into or "infiltrate" the ground. This results in larger amounts of water moving more quickly across a site which can carry more sediment and other pollutants to streams and rivers.

EPA's general permit requires that all industries identify in their Storm Water Pollution Prevention Plans areas that may have a high potential for soil erosion. This includes areas with such heavy activity that plants cannot grow, soil stockpiles, stream banks, steep slopes, construction areas, demolition areas, and any area where the soil is disturbed, denuded (stripped of plants), and subject to wind and water erosion. EPA further requires that you take steps to limit this erosion.

There are seven ways to limit and control sediment and erosion:

1. Leave as much vegetation (plants) on site as possible.

2. Minimize the time that soil is exposed.

3. Prevent runoff from flowing across disturbed areas (divert the flow to vegetated areas).

4. Stabilizing the disturbed soils as soon as possible.

5. Slow down the runoff flowing across the site.

6. Provide drainage ways for the increased runoff (use grassy swales rather than concrete drains).

7. Remove sediment from storm water runoff before it leaves the site.

Using these measures to control erosion and sedimentation is an important part of storm water management. Selecting the best set of sediment and erosion prevention measures for your industry depends upon the nature of the activities on your site (i.e., how much construction or land disturbance there is) and other site-specific conditions (soil type, topography, climate, and season).

In some arid regions, growing vegetation to prevent erosion may be difficult. The local Soil Conservation Service Office or County Extension Office can provide information on any special measures necessary to promote the establishment of vegetation.

Preserving existing vegetation or revegetating disturbed soil as soon as possible after construction is the most effective way to control erosion. A vegetation cover reduces erosion potential in four ways:

1. By shielding the soil surface from direct erosive impact of raindrops;

2. By improving the soil's water storage porosity and capacity so more water can infiltrate into the ground

3. By slowing the runoff and allowing the sediment to drop out or deposit; and

4. By physically holding the soil in place with plant roots.

Vegetative cover can be grass, trees, shrubs, bark, mulch, or straw. Grasses are the most common type of cover used for revegetation because they grow quickly, providing erosion protection within days. Other soil-stabilization practices, such as straw or mulch, may be used during nongrowing seasons to prevent erosion. Newly planted shrubs and trees establish root systems more slowly, so keeping existing ones is a more effective practice.

Vegetative and other site-stabilization practices can be either temporary or permanent controls. Temporary controls provide a cover for exposed or disturbed areas for short periods of time or until permanent erosion controls are put in place. Permanent vegetative practices are used when activities that disturb the soil are completed or when erosion is occurring on a site that is otherwise stabilized. The remainder of this section describes the common vegetative practices listed below:

* Preservation of natural vegetation

* Buffer zones

* Stream bank stabilization

* Mulching, matting, and netting

* Temporary seeding

* Permanent seeding and planting

* Sodding

* Chemical stabilization

9.2.1 Preservation of Natural Vegetation

The preservation of natural vegetation (existing trees, vines, brushes, and grasses) provides natural buffer zones. By preserving stabilized areas, it minimizes erosion potential, protects water quality, and provides aesthetic benefits. This practice is used as a permanent control measure.

This technique is applicable to all types of sites. Areas where preserving vegetation can be particularly beneficial are floodplains, wetlands, stream banks, steep slopes, and other areas where erosion controls would be difficult to establish, install, or maintain.

Preservation of vegetation on a site should be planned before any site disturbance begins. Preservation requires good site management to minimize the impact of construction activities on existing vegetation. Clearly mark the trees to be preserved and protect them from ground disturbances around the base of the tree. Proper maintenance is important to ensure healthy vegetation that can control erosion. Different species, soil types, and climatic conditions will require different maintenance activities, such as mowing, fertilizing, liming, irrigation, pruning, and weed and pest control. Some state/local regulations require natural vegetation to be preserved in sensitive areas; therefore, it is important to consult with the appropriate State/local agencies for more information on their regulations. Maintenance should be performed regularly, especially during construction.

Advantages of Preservation of Natural Vegetation:

* Can handle higher quantities of storm water runoff than newly seeded areas.

* Does not require time to establish (i.e., effective immediately).

* Increases the filtering capacity because the vegetation and root structure are usually denser in preserved natural vegetation than in newly seeded or base areas.

* Enhances aesthetics.

* Provides areas for infiltration, reducing the quantity and velocity of storm water runoff.

* Allows areas where wildlife can remain undisturbed.

* Provides noise buffers and screens for on-site operations.

* Usually requires less maintenance (e.g., irrigation, fertilizer) than planting new vegetation.

Disadvantages of Preservation of Natural Vegetation:

* Requires planning to preserve and maintain the existing vegetation.

* May not be cost effective with high land costs.

* May constrict area available for construction activities.

9.2.2 Buffer Zones

Buffer zones are vegetated strips of land used for temporary or permanent water quality benefits. Buffer zones are used to decrease the velocity of storm water runoff, which in turn helps to prevent soil erosion. Buffer zones are different from vegetated filter strips (see section on Vegetated Filter Strips below), because buffer zone effectiveness is not measured by its ability to improve infiltration (allow water to go into the ground). The buffer zone can be an area of vegetation that is left undisturbed during construction, or it can be newly planted.

Buffer zones can be used at any site that can support vegetation. Buffer zones are particularly effective on floodplains, next to wetlands, along stream banks, and on steep, unstable slopes.

If buffer zones are preserved, existing vegetation, good planning, and site management are needed to protect against disturbances such as grade changes, excavation, damage from equipment, and other activities. Establishing new buffer strips requires the establishment of a good dense turf, trees, and shrubs (see section on Permanent Seeding and Planting below). Careful maintenance is important to ensure healthy vegetation. The need for routine maintenance, such as mowing, fertilizing, liming, irrigating, pruning, and weed and pest control, will depend on the species of plants and trees involved, soil types, and climatic conditions. Maintaining planted areas may require debris removal and protection against unintended uses or traffic. Many state/local storm water programs or zoning authorities have regulations that define required or allowable buffer zones, especially near sensitive areas such as wetlands. Contact the appropriate State/local agencies for their requirements.

Advantages of Buffer Zones:

* Provide aesthetic and water quality benefits.

* Provide areas for infiltration, which reduces amount and speed of storm water runoff.

* Provide areas for wildlife habitat.

* Provide areas for recreation.

* Provide buffers and screens for on-site noise if trees or large bushes are used.

* Low maintenance requirements.

* Low cost when using existing vegetation.

Disadvantages of Buffer Zones:

* May not be cost effective to use if the cost of land is high.

* Are not feasible if land is not available.

* Require plant growth before they are effective.

9.2.3 Stream Bank Stabilization

Stream bank stabilization is used to prevent stream bank erosion from high velocities and quantities of storm water runoff. Typical methods include the following:

* *Riprap:* Large angular stones placed along the stream bank or lake.

* *Gabion:* Rock-filled wire cages that are used to create a new stream bank.

* *Reinforced Concrete:* Concrete bulkheads and retaining walls that replace natural stream banks and create a nonerosive surface.

* *Log Cribbing:* Retaining walls built of logs to anchor the soils against erosive forces. Usually built on the outside of stream bends.

* *Grid Pavers:* Precast or poured-in-place concrete units that are placed along stream banks to stabilize the stream bank and create open spaces where vegetation can be established.

* *Asphalt:* Asphalt paving that is placed along the natural stream bank to create a nonerosive surface.

Stream bank stabilization is used where vegetative stabilization practices are not practical and where the stream banks are subject to heavy erosion from increased flows or disturbance during construction. Stabilization should occur before any land development in the watershed area. Stabilization can also be retrofitted when erosion of a stream bank occurs.

Stream bank stabilization structures should be planned and designed by a professional engineer licensed in the State where the site is located. Applicable federal, state, and local requirements should be followed, including Clean Water Act Section 404 regulations.

An important design feature of stream bank stabilization methods is the foundation of the structure; the potential for the stream to erode the sides and bottom of the channel should be considered to make sure the stabilization measure will be supported properly. Structures can be designed to protect and improve natural wildlife habitats. For example, log structures and grid pavers can be designed to keep vegetation. Only pressure-treated wood should be used in log structures. Permanent structures should be designed to handle expected flood conditions. A well-designed layer of stone can be used in many ways and in many locations to control erosion and sedimentation.

Riprap protects soil from erosion and is often used on steep slopes built with fill materials that are subject to harsh weather or seepage. Riprap can also be used for flow channel liners, inlet and outlet protection at culverts, stream bank protection, and protection of shore lines subject to wave action. It is used where water is turbulent and fast flowing and where soil may erode under the design flow conditions. It is used to expose the water to air as well as to reduce water energy. Riprap and gabion (wire mesh cages filled with rock) are usually placed over a filter blanket (i.e., a gravel layer or filter cloth). Riprap is either a uniform size or graded (different sizes) and is usually applied in an even layer throughout the stream.

Reinforced concrete structures may require positive drainage behind the bulkhead or retaining wall to prevent erosion around the structure. Gabion and grid pavers should be installed according to manufacturers' recommendations.

Stream bank stabilization structures should be inspected regularly and after each large storm event. Structures should be maintained as installed. Structural damage should be repaired as soon as possible to prevent further damage or erosion to the stream bank.

Advantages of Stream Bank Stabilization:

* Can provide control against erosive forces caused by the increase in storm water flows created during land development.

* Usually will not require as much maintenance as vegetative erosion controls.

* May provide wildlife habitats.

* Forms a dense, flexible, self-healing cover that will adapt well to uneven surfaces (riprap).

Disadvantages of Stream Bank Stabilization:

* Does not provide the water quality or aesthetic benefits that vegetative practices could.

* Should be designed by qualified professional engineers, which may increase project costs.

* May be expensive (materials costs).

* May require additional permits for structure.

* May alter stream dynamics which cause changes in the channel downstream.

* May cause negative impacts to wildlife habitats.

9.2.4 Mulching, Matting, and Netting

Mulching is a temporary soil-stabilization or erosion control practice where materials such as grass, hay, woodchips, wood fibers, straw, or gravel are placed on the soil surface. In addition to stabilizing soils, mulching can reduce the speed of storm water runoff over an area. When used together with seeding or planting, mulching can aid in plant growth by holding the seeds, fertilizers, and topsoil in place, by preventing birds from eating seeds, helping to retain moisture, and by insulating against extreme temperatures. Mulch mattings are materials (jute or other wood fibers) that have been formed into sheets of mulch that are more stable than normal mulch. Netting is typically made from jute, other wood fiber, plastic, paper, or cotton and can be used to hold the mulching and matting to the ground. Netting can also be used alone to stabilize soils while the plants are growing; however, it does not retain moisture or temperature well. Mulch binders (either asphalt or synthetic) are sometimes used instead of netting to hold loose mulches together.

Mulching is often used alone in areas where temporary seeding cannot be used because of the season or climate. Mulching can provide immediate, effective, and inexpensive erosion control. On steep slopes and critical areas such as waterways, mulch matting is used with netting or anchoring to hold it in place.

Mulch seeded and planted areas where slopes are steeper than 2:1, where runoff is flowing across the area, or when seedlings need protection from bad weather.

Use of mulch may or may not require a binder, netting, or the tacking of mulch to the ground. Effective netting and matting require firm, continuous contact between the materials and the soil. If there is no contact, the material will not hold the soil and erosion will occur underneath the material. Final grading is not necessary before mulching. Mulched areas should be inspected often to find where mulched material has been loosened or removed. Such areas should be reseeded (if necessary) and the mulch cover replaced immediately. Mulch binders should be applied at rates recommended by the manufacturer or, if asphalt is used, at rates of approximately 480 gallons per acre.

Advantages of Mulching, Matting, and Netting:

* Provide immediate protection to soils that are exposed and that are subject to heavy erosion.

* Retain moisture, which may minimize the need for watering.

* Require no removal because of natural deterioration of mulching and matting.

Disadvantages of Mulching, Matting. and Netting:

* May delay germination of some seeds because cover reduces the soil surface temperature.

* Netting should be removed after usefulness is finished, then landfilled or composted.

9.2.5 Temporary Seeding

Temporary seeding means growing a short-term vegetative cover (plants) on disturbed site areas that may be in danger of erosion. The purpose of temporary seeding is to reduce erosion and sedimentation by stabilizing disturbed areas that will not be stabilized for long periods of time or where permanent plant growth is not necessary or appropriate. This practice uses fast-growing grasses whose root systems hold down the soils so that they are less apt to be carried off site by storm water runoff or wind. Temporary seeding also reduces the problems associated with mud and dust from bare soil surfaces during construction.

Temporary seeding should be performed on areas which have been disturbed by construction and which are likely to be redisturbed, but not for several weeks or more. Typical areas might include denuded areas, soil stockpiles, dikes, dams, sides of sediment basins, and temporary roadbanks. Temporary seeding should take place as soon as practicable after the last land-disturbing activity in an area. Check the requirements of your permit for the maximum amount of time allowed between the last disturbance of an area and temporary stabilization.

Temporary seeding may not be an effective practice in arid and semiarid regions where the climate prevents fast plant growth, particularly during the dry seasons. In those areas, mulching or chemical stabilization may be better for the shortterm (see sections on Mulching and Chemical Stabilization).

Proper seed bed preparation and the use of high-quality seed are needed to grow plants for effective erosion control. Soil that has been compacted by heavy traffic or machinery may need to be loosened. Successful growth usually requires that the soil be tilled before the seed is applied. Topsoiling is not necessary for temporary seeding; however, it may improve the chances of establishing temporary vegetation in an area. Seed bed preparation may also require applying fertilizer and/or lime to the soil to make conditions more suitable for plant growth. Proper fertilizer, seeding mixtures, and seeding rates vary depending on the location of the site, soil types, slopes, and season. Local suppliers, State and local regulatory agencies, and the USDA Soil Conservation Service will supply information on the best seed mixes and soil conditioning methods.

Seeded areas should be covered with mulch to provide protection from the weather. Seeding on slopes of 2:1 or more, in adverse soil conditions, during excessively hot or dry weather, or where heavy rain is expected should be followed by spreading mulch (see section on Mulching). Frequent inspections are necessary to check that conditions for growth are good. If the plants do not grow quickly or thick enough to prevent erosion, the area should be reseeded as soon as possible. Seeded areas should be kept adequately moist. If normal rainfall will not be enough, mulching, matting, and controlled watering should be done. If seeded areas are watered, watering rates should be watched so that over-irrigation (which can cause erosion itself) does not occur.

Advantages of Temporary Seeding:

- Is generally inexpensive and easy to do.
- Establishes plant cover fast when conditions are good.
- Stabilizes soils well, is aesthetic, and can provide sedimentation controls for other site areas.
- May help reduce costs of maintenance on other erosion controls (e.g., sediment basins may need to be cleaned out less often).

Disadvantages of Temporary Seeding:

- Depends heavily on the season and rainfall rate for success.

* May require extensive fertilizing of plants grown on some soils, which can cause problems with local water quality.

* Requires protection from heavy use, once seeded.

* May produce vegetation that requires irrigation and maintenance.

9.2.6 Permanent Seeding and Planting

Permanent seeding of grass and planting trees and brush provide stabilization to the soil by holding soil particles in place. Vegetation reduces sediments and runoff to downstream areas by slowing the velocity of runoff and permitting greater infiltration of the runoff. Vegetation also filters sediments, helps the soil absorb water, improves wildlife habitats, and enhances the aesthetics of a site.

Permanent seeding and planting is appropriate for any graded or cleared area where long-lived plant cover is desired. Some areas where permanent seeding is especially important are filter strips, buffer areas, vegetated swales, steep slopes, and stream banks. This practice is effective on areas where soils are unstable because of their texture, structure, a high water table, high winds, or high slope. When seeding in northern areas during fall or winter, cover the area with mulch to provide a protective barrier against cold weather (see Mulching). Seeding should also be mulched if the seeded area slopes 4:1 or more, if soil is sandy or clayey, or if weather is excessively hot or dry. Plant when conditions are most favorable for growth. When possible, use low-maintenance local plant species. Install all other erosion control practices such as dikes, basins, and surface runoff control measures before planting.

For this practice to work, it is important to select appropriate vegetation, prepare a good seedbed, properly time planting, and water and fertilize. Planting local plants during their regular growing season will increase the chances for success and may lessen the need for watering. Check seeded areas frequently for proper watering and growth conditions.

Topsoil should be used on areas where topsoils have been removed, where the soils are dense or impermeable, or where mulching and fertilizers alone cannot improve soil quality. Topsoiling should be coordinated with the seeding and planting practices and should not be planned while the ground is frozen or too wet. Topsoil layers should be at least 2 inches deep (or similar to the existing topsoil depth).

To minimize erosion and sedimentation, remove as little existing topsoil as possible. All site controls should be in place before the topsoil is removed. If topsoils are brought in from another site, it is important that its texture is compatible with the subsoils on site; for example, sandy topsoils are not compatible with clay subsoils.

Stockpiling of topsoils on site requires good planning so that soils will not obstruct other operations. If soil is to be stockpiled, consider using temporary seeding, mulching, or silt fencing to prevent or control erosion. Inspect the stockpiles frequently for erosion. After topsoil has been spread, inspect it regularly, and reseed or replace areas that have eroded.

Advantages of Permanent Seeding and Planting:

* Improve the aesthetics of a site.
* Provide excellent stabilization.
* Provide filtering of sediments.
* Provide wildlife habitat.
* Are relatively inexpensive.

Disadvantages of Permanent Seeding and Planting:

* May require irrigation to establish vegetation.
* Depend initially on climate and weather for success.

9.2.7 Sodding

Sodding stabilizes an area by establishing permanent vegetation, providing erosion and sedimentation controls, and providing areas where storm water can infiltrate the ground.

Sodding is appropriate for any graded or cleared area that might erode and where a permanent, long-lived plant cover is needed immediately. Examples of where sodding can be used are buffer zones, stream banks, dikes, swales, slopes, outlets, level spreaders, and filter strips.

The soil surface should be fine graded before laying down the sod. Topsoil may be needed in areas where the soil textures are inadequate (see topsoil discussion in section on Permanent Seeding and Planting). Lime and fertilizers should be added to the soil to promote good growth conditions. Sodding can be applied in alternating strips or other patterns, or alternate areas can be seeded to reduce expense. Sod should not be planted during very hot or wet weather. Sod should not be placed on slopes that are greater than 3:1 if they are to be mowed. If placed on steep slopes, sod should be laid with staggered joints and/or be pegged. In areas such as steep slopes or next to running waterways, chicken wire, jute, or other netting can be placed over the sod for extra protection against lifting (see Mulching, Matting, and Netting). Roll or compact immediately after installation to ensure firm contact with the underlying topsoil. Inspect the sod frequently after it is first installed, especially after large storm events, until it is established as permanent cover. Remove and replace dead sod. Watering may be necessary after planting and during periods of intense heat and/or lack of rain.

Advantages of Sodding:

- Can provide immediate vegetative cover and erosion control.

- Provides more stabilizing protection than initial seeding through dense cover formed by sod.

- Produces lower weed growth than seeded vegetation.

- Can be used for site activities within a shorter time than can seeded vegetation.

* Can be placed at any time of the year as long as moisture conditions in the soil are favorable, except when the ground is frozen.

Disadvantages of Sodding:

* Purchase and installation costs are higher than for seeding.

* May require continued irrigation if the sod is placed during dry seasons or on sandy soils.

9.2.8 Chemical Stabilization

Chemical stabilization practices, often referred to as a chemical mulch, soil binder, or soil palliative, are temporary erosion control practices. Materials made of vinyl, asphalt, or rubber are sprayed onto the surface of the soil to hold the soil in place and protect against erosion from storm water runoff and wind. Many of the products used for chemical stabilization are human made, and many different products are on the market.

Chemical stabilization can be used as an alternative in areas where temporary seeding practices cannot be used because of the season or climate. It can provide immediate, effective, and inexpensive erosion control anywhere erosion is occurring on a site.

The application rates and procedures recommended by the manufacturer of a chemical stabilization product should be followed as closely as possible to prevent the products from forming ponds and from creating large areas where moisture cannot get through.

Advantages of Chemical Stabilization:

* Is easily applied to the surface of the soil.

* Is effective in stabilizing areas where plants will not grow.

* Provides immediate protection to soils that are in danger of erosion.

Disadvantages of Chemical Stabilization:

* Can create impervious surfaces (where water cannot get through), which may in turn increase the amount and speed of storm water runoff.

* May cause harmful effects on water quality if not used correctly. Is usually more expensive than vegetative cover.

9.3 STRUCTURAL EROSION PREVENTION AND SEDIMENT CONTROL PRACTICES

Structural practices used in sediment and erosion control divert storm water flows away from exposed areas, convey runoff, prevent sediments from moving off site, and can also reduce the erosive forces of runoff waters. The controls can either be used as permanent or temporary measures. Practices discussed include the following:

* Interceptor dikes and swales

* Pipe slope drains

* Subsurface drains

* Silt fence

* Straw bale barrier

* Brush barrier

* Gravel or stone filter berm

* Storm drain inlet protection

* Sediment trap

* Temporary sediment basin

* Outlet protection

* Check dams

* Surface roughening

* Gradient terraces

9.3.1 Interceptor Dikes and Swales

Interceptor dikes (ridges of compacted soil) and swales (excavated depressions) are used to keep upslope runoff from crossing areas where there is a high risk of erosion. They reduce the amount and speed of flow and then guide it to a stabilized outfall (point of discharge) or sediment trapping area. Interceptor dikes and swales divert runoff using a combination of earth dike and vegetated swale. Runoff is channeled away from locations where there is a high risk of erosion by placing a diversion dike or swale at the top of a sloping disturbed area. Dikes and swales also collect overland flow, changing it into concentrated flows (i.e., flows that are combined). Interceptor dikes and swales can be either temporary or permanent storm water control structures.

Interceptor dikes and swales are generally built around the perimeter of a construction site before any major soil-disturbing activity takes place. Temporary dikes or swales may also be used to protect existing buildings; areas, such as stockpiles; or other small areas that have not yet been fully stabilized. When constructed along the upslope perimeter of a disturbed or high-risk area (though not necessarily all the way around it), dikes or swales prevent runoff from uphill areas from crossing the unprotected slope. Temporary dikes or swales constructed on the downslope side of the disturbed or high-risk area will prevent runoff that contains sediment from leaving the site before sediment is removed. For short slopes, a dike or swale at the top of the slope reduces the amount of runoff reaching the disturbed area. For longer slopes, several dikes or swales are placed across the slope at intervals. This practice reduces the amount of runoff that accumulates on the face of the slope and carries the runoff safely down the slope. In all cases, runoff is guided to a sediment trapping area or a stabilized outfall before release.

Temporary dikes and swales are used in areas of overland flow; if they remain in place longer than 15 days, they should be stabilized. Runoff channeled by a dike or swale should be directed to an adequate sediment trapping area or stabilized outfall. Care should be taken to provide enough slope for drainage, but not too much slope to cause erosion due to high runoff flow speed. Temporary interceptor dikes and swales may remain in place as long as 12 to 18 months (with proper stabilization) or be rebuilt at the end of each day's activities. Dikes or swales should remain in place until the area they were built to protect is permanently stabilized. Interceptor dikes and swales can be permanent controls. However, permanent controls should be designed to handle runoff after construction is complete; should be permanently stabilized; and should be inspected and maintained on a regular basis. Temporary and permanent control measures should be inspected once each week on a regular schedule and after every storm. Repairs necessary to the dike and flow channel should be made promptly.

Advantages of Interceptor Dikes and Swales:

* Are simple and effective for channeling runoff away from areas subject to erosion.

* Can handle flows from large drainage areas.

* Are inexpensive because they use materials and equipment normally found on site.

Disadvantages of Interceptor Dikes and Swales:

* If constructed improperly, can cause erosion and sediment transport, since flows are concentrated.

* May cause problems to vegetation growth if water flow is too fast.

* Require additional maintenance, inspections, and repairs.

9.3.2 Pipe Slope Drains

Pipe slope drains reduce the risk of erosion by discharging runoff to stabilized areas. Made of flexible or rigid pipe, they carry concentrated runoff from the top to the bottom of a slope that has already been damaged by erosion or is at high risk for erosion. They are also used to drain saturated slopes that have the potential for soil slides. Pipe slope drains can be either temporary or permanent, depending on the method of installation and material used.

Pipe slope drains are used whenever it is necessary to convey water down a slope without causing erosion. They are especially effective before a slope has been stabilized or before permanent drainage structures are ready for use. Pipe slope drains may be used with other devices, including diversion dikes or swales, sediment traps, and level spreaders (used to spread out storm water runoff uniformly over the surface of the ground). Temporary pipe slope drains, usually flexible tubing or conduit, may be installed prior to the construction of permanent drainage structures. Permanent slope drains may be placed on or beneath the ground surface; pipes, sectional downdrains, paved chutes, or clay tiles may be used.

Paved chutes may be covered with a surface of concrete or other impenetrable material. Subsurface drains can be constructed of concrete, PVC, clay tile, corrugated metal, or other permanent material.

The drain design should be able to handle the volume of flow. The effective life span of a temporary pipe slope drain is up to 30 days after permanent stabilization has been achieved. The maximum recommended drainage area for pipe slope drains is 10 acres.

The inlets and outlets of a pipe slope drain should be stabilized. This means that a flared end section should be used at the entrance of the pipe. The soil around the pipe entrance should be fully compacted. The soil at the discharge end of the pipe should be stabilized with riprap (a combination of large stones, cobbles, and boulders). The riprap should be placed along the bottom of a swale which leads to a sediment-trapping structure or another stabilized area.

Pipe slope drains should be inspected on a regular schedule and after any major storm. Be sure that the inlet from the pipe is properly installed to prevent bypassing the inlet and undercutting the structure. If necessary, install a headwall, riprap, or sandbags around the inlet. Check the outlet point for erosion and check the pipe for breaks or clogs. Install outlet protection if needed and promptly clear breaks and clogs.

Advantages of Pipe Slope Drains:

* Can reduce or eliminate erosion by transporting runoff down steep slopes or by draining saturated soils.

* Are easy to install and require little maintenance.

Disadvantages of Pipe Slope Drains:

* Require that the area disturbed by the installation of the drain should be stabilized or it, too, will be subject to erosion.

* May clog during a large storm.

9.3.3 Subsurface Drains

A subsurface drain is a perforated pipe or conduit placed beneath the surface of the ground at a designed depth and grade. It is used to drain an area by lowering the water table. A high water table can saturate soils and prevent the growth of certain types of vegetation. Saturated soils on slopes will sometimes "slip" down the hill. Installing subsurface drains can help prevent these problems.

There are two types of subsurface drains: relief drains and interceptor drains. Relief drains are used to dewater an area where the water table is high. They may be placed in a gridiron, herringbone, or random pattern. Interceptor drains are used to remove water where sloping soils are excessively wet or subject to slippage. They are usually placed as single pipes instead of in patterns. Generally, subsurface drains are suitable only in areas where the soil is deep enough for proper installation. They are not recommended where they pass under heavy-vehicle crossings.

Drains should be placed so that tree roots will not interfere with drainage pipes. The drain design should be adequate to handle the volume of flow. Areas disturbed by the installation of a drain should be stabilized or they, too, will be subject to erosion. The soil layer must be deep enough to allow proper installation.

Backfill immediately after the pipe is placed. Material used for backfill should be open granular soil that is highly permeable. The outlet should be stabilized and should direct sediment-laden storm water runoff to a sediment trapping structure or another stabilized area.

Inspect subsurface drains on a regular schedule and check for evidence of pipe breaks or clogging by sediment, debris, or tree roots. Remove blockage immediately, replace any broken sections, and restabilize the surface. If the blockage is from tree roots, it may be necessary to relocate the drain. Check inlets and outlets for sediment or debris. Remove and dispose of these materials properly.

Advantages of Subsurface Drains:

* Provide an effective method for stabilizing wet sloping soils.

* Are an effective way to lower the water table.

Disadvantages of Subsurface Drains:

* May be pierced and clogged by tree roots.

* Should not be installed under heavy-vehicle crossings.

* Cost more than surface drains because of the expenses of excavation for installation.

9.3.4 Silt Fence

A silt fence, also called a "filter fence," is a temporary measure for sedimentation control. It usually consists of posts with filter fabric stretched across the posts, and sometimes with a wire support fence. The lower edge of the fence is vertically trenched and covered by backfill. A silt fence is used in small drainage areas to detain sediment. These fences are most effective where there is overland flow (runoff that flows over the surface of the ground as a thin, even layer) or in minor swales or drainageways. They prevent sediment from entering receiving waters. Silt fences are also used to catch wind-blown sand and to create an anchor for sand dune creation. Aside from the traditional wooden post and filter fabric method, there are several variations of silt fence installation, including silt fences that can be purchased with pockets presewn to accept use of steel fence posts.

A silt fence should be installed prior to major soil disturbance in the drainage area. Such a structure is only appropriate for drainage areas of 1 acre or less with velocities of 0.5 cfs or less. The fence should be placed across the bottom of a slope or minor drainageway along a line of uniform elevation (perpendicular to the direction of flow). It can be used at the outer boundary of the work area. However, the fence does not have to surround the work area completely. In addition, a silt fence is effective where sheet and rill erosion may be a problem. Silt fences should not be constructed in streams or swales.

A silt fence is not appropriate for a large area or where the flow rate is greater than 0.5 cfs. This type of fence can be more effective than a straw bale barrier if properly installed and maintained. It may be used in combination with other erosion and sediment practices.

The effective life span for a silt fence is approximately 6 months. During this period, the fence requires frequent inspection and prompt maintenance to maintain its effectiveness. Inspect the fence after each rainfall. Check for areas where runoff may have eroded a channel beneath the fence, or where the fence was caused to sag or collapse by runoff flowing over the top. Remove and properly dispose of sediment when it is one-third to one-half the height of the fence or after each storm.

Advantages of a Silt Fence:

* Removes sediments and prevents downstream damage from sediment deposits.

* Reduces the speed of runoff flow.

* Minimal clearing and grubbing required for installation.

* Inexpensive.

Disadvantages of a Silt Fence:

* May result in failure from improper choice of pore size in the filter fabric or improper installation.

* Should not be used in streams.

* Is only appropriate for small drainage areas with overland flow.

* Frequent inspection and maintenance are necessary to ensure effectiveness.

9.3.5 Straw Bale Barrier

Straw bales can be used as a temporary sediment barrier. They are placed end to end in a shallow excavated trench (with no gaps in between) and staked into place. If properly installed, they can detain sediment and reduce flow velocity from small drainage areas. A straw bale barrier prevents sediment from leaving the site by trapping the sediment in the barrier while allowing the runoff to pass through. It can also be used to decrease the velocity of sheetflow or channel flows of low to moderate levels.

A straw bale barrier should be installed prior to major soil disturbance in the drainage area. This type of barrier is placed perpendicular to the flow, across the bottom of a slope or minor drainageway where there is sheetflow. It can be used at the perimeter of the work area, although is does not have to surround it completely. It can also be very effective when used in combination with other erosion and sediment control practices. A straw bale barrier may be used where the length of slope behind the barrier is less than 100 feet and where the slope is less than 2:1.

The success of a straw bale barrier depends on proper installation. The bales must be firmly staked into the entrenchment and the entrenchment must be properly backfilled. To function effectively, the bales must be placed end to end and there can be no gaps between the bales.

Straw bale barriers are useful for approximately 3 months. They must be inspected and repaired immediately after each rainfall or daily if there is prolonged rainfall. Damaged straw bales require immediate replacement. After each storm, or on a regular basis, trapped sediments must be removed and disposed of properly.

Advantages of a Straw Bale Barrier:

* Can prevent downstream damage from sediment deposits if properly installed, used, and maintained.

* Can be an inexpensive way to reduce or prevent erosion.

Disadvantages of a Straw Bale Barrier:

* May not be used in streams or large swales.

* Poses a risk of washouts if the barrier is installed improperly or a storm is severe.

* Has a short life span and a high inspection and maintenance requirement.

* Is appropriate for only small drainage areas.

* Is easily subject to misuse and can contribute to sediment problems.

9.3.6 Brush Barrier

A brush barrier is a temporary sediment barrier constructed from materials resulting from on-site clearing and grubbing. It is usually constructed at the bottom perimeter of the disturbed area. Filter fabric is sometimes used as an anchor over the barrier to increase its filtering efficiency. Brush barriers are used to trap and retain small amounts of sediment by intercepting the flow from small areas of soil disturbance.

A brush barrier should only be used to trap sediment from runoff coming from a small drainage area. The slope that the brush barrier is placed across should be very gentle. Do not place a brush barrier in a swale or any other channel. Brush barriers should be constructed below areas subject to erosion.

The construction of a brush barrier should be started as soon as clearing and grubbing has produced enough material to make the structure. Wood chips should not be included in the material used for the barrier because of the possibility of leaching. When the site has been stabilized and any excess sediment has been disposed of properly, the filter fabric can be removed. Over time, natural vegetation will establish itself within the barrier, and the barrier itself will decompose.

You will not have to maintain the brush barrier unless there is a very large amount of sediment being deposited. If used, the filter fabric anchor should be checked for tears and the damaged sections replaced promptly. The barrier should be inspected after each rainfall and checked for areas breached by concentrated flow. If necessary, repairs should be made promptly and excess sediment removed and disposed of properly.

Advantages of a Brush Barrier:

* Can help prevent downstream damage from sediment deposits.

* Is constructed of cleared on-site materials and, thus, is inexpensive.

* Usually requires little maintenance, unless there are very heavy sediment deposits.

Disadvantages of a Brush Barrier:

* Does not replace a sediment trap or basin.

* Is only appropriate for small drainage areas.

* Has very limited sediment retention.

9.3.7 Gravel or Stone Filter Berm

A gravel or stone filter berm is a temporary ridge constructed of loose gravel, stone, or crushed rock. It slows and filters flow, diverting it from an exposed traffic area. Diversions constructed of compacted soil may be used where there will be little or no construction traffic within the right-of-way. They are also used for directing runoff from the right-of-way to a stabilized outlet.

This method is appropriate where roads and other rights-of-way under construction should accommodate vehicular traffic. Berms are meant for use in areas with shallow slopes. They may also be used at traffic areas within the construction site.

Berm material should be well-graded gravel or crushed rock. The spacing of the berms will depend on the steepness of the slope: berms should be placed closer together as the slope increases. The diversion should be inspected daily, after each rainfall, or if breached by construction or other vehicles. All needed repairs should be performed immediately. Accumulated sediment should be removed and properly disposed of and the filter material replaced, as necessary.

Advantages of a Gravel or Stone Filter Berm:

* Is a very efficient method of sediment control.

Disadvantages of a Gravel or Stone Filter Berm:

* Is more expensive than methods that use on-site materials.

* Has a very limited life span.

* Can be difficult to maintain because of clogging from mud and soil on vehicle tires.

9.3.8 Storm Drain Inlet Protection

Storm drain inlet protection is a filtering measure placed around any inlet or drain to trap sediment. This mechanism prevents the sediment from entering inlet structures. Additionally, it serves to prevent the silting-in of inlets, storm drainage systems, or receiving channels. Inlet protection may be composed of gravel and stone with a wire mesh filter, block and gravel, filter fabric, or sod.

This type of protection is appropriate for small drainage areas where storm drain inlets will be ready for use before final stabilization. Storm drain inlet protection is also used where a permanent storm drain structure is being constructed on site. Straw bales are not recommended for this purpose. Filter fabric is used for inlet protection when storm water flows are relatively small with low velocities. This practice cannot be used where inlets are paved because the filter fabric should be staked. Block and gravel filters can be used where velocities are higher. Gravel and mesh filters can be used where flows are higher and subject to disturbance by site traffic. Sod inlet filters are generally used where sediments in the storm water runoff are low.

Storm drain inlet protection is not meant for use in drainage areas exceeding 1 acre or for large concentrated storm water flows. Installation of this measure should take place before any soil disturbance in the drainage area. The type of material used will depend on site conditions and the size of the drainage area. Inlet protection should be used in combination with other measures, such as small impoundments or sediment traps, to provide more effective sediment removal. Inlet protection structures should be inspected regularly, especially after a rainstorm. Repairs and silt removal should be performed as necessary. Storm drain inlet protection structures should be removed only after the disturbed areas are completely stabilized.

Advantages of Storm Drain Inlet Protection:

* Prevents clogging of existing storm drainage systems and the siltation of receiving waters.

* Reduces the amount of sediment leaving the site.

Disadvantages of Storm Drain Inlet Protection:

* May be difficult to remove collected sediment.

* May cause erosion elsewhere if clogging occurs.

* Is practical only for low-sediment, low-volume flows.

9.3.9 Sediment Trap

A sediment trap is formed by excavating a pond or by placing an earthen embankment across a low area or drainage swale. An outlet or spillway is constructed using large stones or aggregate to slow the release of runoff. The trap retains the runoff long enough to allow most of the silt to settle out.

A temporary sediment trap may be used in conjunction with other temporary measures, such as gravel construction entrances, vehicle wash areas, slope drains, diversion dikes and swales, or diversion channels. This device is appropriate for sites with short time schedules.

Sediment traps are suitable for small drainage areas, usually no more than 10 acres, that have no unusual drainage features. The trap should be large enough to allow the sediments to settle and should have a capacity to store the collected sediment until it is removed. The volume of storage required depends upon the amount and intensity of expected rainfall and on estimated quantities of sediment in the storm water runoff. Check your permit to see if it specifies a minimum storage volume for sediment traps.

A sediment trap is effective for approximately 18 months. During this period, the trap should be readily accessible for periodic maintenance and sediment removal. Traps should be inspected after each rainfall and cleaned when no more than half the design volume has been filled with collected sediment. The trap should remain in operation and be properly maintained until the site area is permanently stabilized by vegetation and/or when permanent structures are in place.

Advantages of a Sediment Trap:

* Protects downstream areas from clogging or damage due to sediment deposits.

* Is inexpensive and simple to install.

* Can simplify the design process by trapping sediment at specific spots on site.

Disadvantages of a Sediment Trap:

* Is suitable only for a limited area.

* Is effective only if properly maintained.

* Will not remove very fine silts and clays.

* Has a short life span.

9.3.10 Temporary Sediment Basin

A temporary sediment basin is a settling pond with a controlled storm water release structure used to collect and store sediment produced by construction activities. A sediment basin can be constructed by excavation or by placing an earthen embankment across a low area or drainage swale. Sediment basins can be designed to maintain a permanent pool or to drain completely dry. The basin detains sediment-laden runoff from larger drainage areas long enough to allow most of the sediment to settle out.

The pond has a gravel outlet or spillway to slow the release of runoff and provide some sediment filtration. By removing sediment, the basin helps prevent clogging of off-site conveyance systems and sediment loading of receiving waterways. In this way, the basin helps prevent destruction of waterway habitats.

A temporary sediment basin should be installed before clearing and grading is undertaken. It should not be built on an embankment in an active stream. The creation of a dam in such a site may result in the destruction of aquatic habitats. Dam failure can also result in flooding. A temporary sediment basin should be located only where there is sufficient space and appropriate topography. The basin should be made large enough to handle the maximum expected amount of site drainage. Fencing around the basin may be necessary for safety or vandalism reasons.

A temporary sediment basin used in combination with other control measures, such as seeding or mulching, is especially effective for removing sediments.

Temporary sediment basins are usually designed for disturbed areas larger than 5 acres. The pond should be large enough to hold runoff long enough for sediment to settle. Sufficient space should be allowed for collected sediments. Check the requirements of your permit to see if there is a minimum storage requirement for sediment basins. The useful life of a temporary sediment basin is about 12 to 18 months.

Sediment trapping efficiency is improved by providing the maximum surface area possible. Because liner silts may not settle out completely, additional erosion control measures should be used to minimize release of fine silt. Runoff should enter the basin as far from the outlet as possible to provide maximum retention time.

Sediment basins should be readily accessible for maintenance and sediment removal. They should be inspected after each rainfall and be cleaned out when about half the volume has been filled with sediment. The sediment basin should remain in operation and be properly maintained until the site area is permanently stabilized by vegetation and/or when permanent structures are in place. The embankment forming the sedimentation pool should be well compacted and stabilized with vegetation. If the pond is located near a residential area, it is recommended for safety reasons that a sign be posted and that the area be secured by a fence. A well-built temporary sediment basin that is large enough to handle the post construction runoff volume may later be converted to use as a permanent storm water management structure.

Advantages of a Temporary Sediment Basin:

* Protects downstream areas from clogging or damage due to sediment deposits generated during construction activities.

* Can trap smaller sediment particles than sediment traps can because of the longer detention time.

Disadvantages of a Temporary Sediment Basin:

* Is generally suitable for small areas.

* Requires regular maintenance and cleaning.

* Will not remove very fine silts and clays unless used in conjunction with other measures.

* Is a more expensive way to remove sediment than several other methods.

* Requires careful adherence to safety practices since ponds are attractive to children.

9.3.11 Outlet Protection

Outlet protection reduces the speed of concentrated storm water flows and, therefore, reduces erosion or scouring at storm water outlets and paved channel sections. In addition, outlet protection lowers the potential for downstream erosion. This type of protection can be achieved through a variety of techniques, including stone or riprap, concrete aprons, paved sections, and settling basins installed below the storm drain outlet.

Outlet protection should be installed at all pipe, interceptor dike, swale, or channel section outlets where the velocity of flow may cause erosion at the pipe outlet and in the receiving channel. Outlet protection should also be used at outlets where the velocity of flow at the design capacity may result in plunge pools (small permanent pools located at the inlet to or the outfall from BMPs). Outlet protection should be installed early during construction activities, but may be added at any time, as necessary.

The exit velocity of the runoff as it leaves the outlet protection structure should be reduced to levels that minimize erosion. Outlet protection should be inspected on a regular schedule to look for erosion and scouring. Repairs should be made promptly.

Advantages of Outlet Protection:

* Provides, with riprap-line apron (the most common outlet protection), a relatively low-cost method that can be installed easily on most sites.

* Removes sediment in addition to reducing flow speed.

* Can be used at most outlets where the flow speed is high.

* Is an inexpensive but effective measure.

* Requires less maintenance than many other measures.

Disadvantages of Outlet Protection:

* May be unsightly.

* May cause problems in removing sediment (without removing and replacing the outlet protection structure itself).

* May require frequent maintenance for rock outlets with high-velocity flows.

9.3.12 Check Dams

A check dam is a small, temporary or permanent dam constructed across a drainage ditch, swale, or channel to lower the speed of concentrated flows. Reduced runoff speed reduces erosion and gullying in the channel and allows sediments and other pollutants to settle out.

A check dam should be installed in steeply sloped swales, or in swales where adequate vegetation cannot be established. A check dam may be built from logs, stone, or pea gravel-filled sandbags.

Check dams should be used only in small open channels that drain 10 acres or less. The dams should not be placed in streams (unless approved by appropriate state authorities). The center section of the check dam should be lower than the edges. Dams should be spaced so that the toe of the upstream dam is at the same elevation as the top of the downstream dam.

After each significant rainfall, check dams should be inspected for sediment and debris accumulation. Sediment should be removed when it reaches one half the original dam height. Check for erosion at edges and repair promptly as required. After construction is complete, all stone and riprap should be removed if vegetative erosion controls will be used as a permanent erosion control measure. It will be important to know the expected erosion rates and runoff flow rate for the swale in which this measure is to be installed. Contact the State/local storm water program agency or a licensed engineer for assistance in designing this measure.

Advantages of Check Dams:

* Are inexpensive and easy to install.

* May be used permanently if designed properly.

* Allow a high proportion of sediment in the runoff to settle out.

* Reduce velocity and provide aeration of the water.
* May be used where it is not possible to divert the flow or otherwise stabilize the channel.

Disadvantages of Check Dams:

* May kill grass linings in channels if the water level remains high after it rains or if there is significant sedimentation.

* Are useful only for drainage areas of 10 acres or less.

9.3.13 Surface Roughening

Surface roughening is a temporary erosion control practice. The soil surface is roughened by the creation of horizontal grooves, depressions, or steps that run parallel to the contour of the land. Slopes that are not fine graded and that are left in a toughened condition can also control erosion. Surface roughening reduces the speed of runoff, increases infiltration, and traps sediment. Surface roughening also helps establish vegetative cover by reducing runoff velocity and giving seed an opportunity to take hold and grow.

Surface roughening is appropriate for all slopes. To slow erosion, roughening should be done as soon as possible after the vegetation has been removed from the slope. Roughening can be used with both seeding and planting and temporary mulching to stabilize an area. For steeper slopes and slopes that will be left roughened for longer periods of time, a combination of surface roughening and vegetation is appropriate.

Different methods can be used to roughen the soil surface on slopes. They include stair-step grading, grooving (using disks, spring harrows, or teeth on a front-end loader), and tracking (driving a crawler tractor up and down a slope, leaving the cleat imprints parallel to the slope contour). The selection of an appropriate method depends on the grade of the slope, mowing requirements after vegetative cover is established, whether the slope was formed by cutting or filling, and the type of equipment available.

Cut slopes with a gradient steeper than 3:1, but less than 2:1 should be stair-step graded or groove cut. Stair-step grading works well with soils containing large amounts of small rock. Each step catches material discarded from above and provides a level site where vegetation can grow. Stairs should be wide enough to work with standard earth-moving equipment. Grooving can be done by any implement that can be safely operated on the slope, including those described above. Grooves should not be less than 3 inches deep nor more than 15 inches apart. Fill slopes with a gradient steeper than 3:1, but less than 2:1 should be compacted every 9 inches of depth. The face of the slope should consist of loose, uncompacted fill 4 to 6 inches deep that can be left rough or can be grooved as described above, if necessary.

Any cut or filled slope that will be mowed should have a gradient less than 3:1. Such a slope can be roughened with shallow grooves parallel to the slope contour by using normal tilling. Grooves should be close together (less than 10 inches) and not less than 1 inch deep. Any gradient with a slope greater than 2:1 should be stair-stepped.

It is important to avoid excessive compacting of the soil surface, especially when tracking, because soil compaction inhibits vegetation growth and causes higher runoff speed. Therefore, it is best to limit roughening with tracked machinery to sandy soils that do not compact easily and to avoid tracking on clay soils. Surface-roughened areas should be seeded as quickly as possible. Also, regular inspections should be made of all surface-roughened areas, especially after storms. If rills (small watercourses that have steep sides and are usually only a few inches deep) appear, they should be filled, graded again, and reseeded immediately. Proper dust-control procedures should be followed when surface roughening.

Advantages of Surface Roughening:

* Provides a degree of instant erosion protection for bare soil while vegetative cover is being established.

* Is inexpensive and simple for short-term erosion control.

Disadvantages of Surface Roughening:

* Is of limited effectiveness in anything more than a gentle rain.

* Is only temporary; if roughening or vegetative cover is washed away in a heavy storm or the vegetation does not take hold, the surface will have to be re-roughened and new seed laid.

9.3.14 Gradient Terraces

Gradient terraces are earth embankments or ridge and channels constructed with suitable spacing and with an appropriate grade. They reduce erosion damage by capturing surface runoff and directing it to a stable outlet at a speed that minimizes erosion.

Gradient terraces are usually limited to use on land that has no vegetation and that has a water erosion problem, or where it is anticipated that water erosion will be a problem. Gradient terraces should not be constructed on slopes with sandy or rocky soils. They will be effective only where suitable runoff outlets are or will be made available.

Gradient terraces should be designed and installed according to a plan determined by an engineering survey and layout. It is important that gradient terraces are designed with adequate outlets, such as a grassed waterway, vegetated area, or tile outlet. In all cases, the outlet should direct the runoff from the terrace system to a point where the outflow will not cause erosion or other damage. Vegetative cover should be used in the outlet where possible. The design elevation of the water surface of the terrace should not be lower than the design elevation of the water surface in the outlet at their junction, when both are operating at design flow. Terraces should be inspected regularly at least once a year and after major storms. Proper dust-control procedures should be followed while constructing these features.

Advantages of Gradient Terraces:

* Reduce runoff speed and increase the distance of overland runoff flow.

* Hold moisture better than do smooth slopes and minimize sediment loading of surface runoff.

Disadvantages of Gradient Terraces:

* May significantly increase cut and fill costs and cause sloughing if excessive water infiltrates the soil.

* Are not practical for sandy, steep, or shallow soils.

9.4 INFILTRATION PRACTICES

Infiltration practices are surface or subsurface measures that allow for quick infiltration of storm water runoff. Rapid infiltration is possible because the structures or soils used in these practices are very porous. Infiltration practices offer an advantage over other practices in that they provide some treatment of runoff, preserve the natural flow in streams, and recharge groundwater. Many of the infiltration practices also can reduce the velocity of the runoff so that it will not cause damaging erosion. Another benefit of infiltration practices is that they reduce the need for expensive storm water conveyance systems. Construction and maintenance of these practices may, however, require some level of expertise to prevent clogging and to retain high effectiveness.

Infiltration BMPs are not practical in all cases. These practices should not be used in areas where runoff is contaminated with pollutants other than sediment or oil and grease. Excessively drained (i.e., very sandy) soils may provide inadequate treatment of runoff, which could result in groundwater contamination. Other site-specific conditions, such as depth to bedrock or depth to the water table, could limit their use or make it impossible to use infiltration BMPs. Also, infiltration practices should not be installed near wells, foundations, septic tank drainfields, or on unstable slopes.

The infiltration practices in this section have been divided into two categories: vegetative infiltration practices and infiltration structures. Vegetative infiltration practices rely on vegetated soils that are well drained to provide storage for the infiltration of storm water. Soils used for this practice generally have not previously been disturbed or compacted so that they more easily allow infiltration. Once vegetation has been planted, use of the area must be limited or the practice may not operate efficiently. The practices that are discussed include vegetated filter strips, grassed swales, and level spreaders.

Infiltration structures are built over soils to aid in collection of storm water runoff and are designed to allow storm water to infiltrate into the ground. These structures generally require a level of expertise for both their design and construction so that they function properly. Maintenance activities are very important because infiltration structures are easily damaged by high sediment loads. Often, infiltration structures are used with other structures that pretreat the storm water runoff for sediments, oil, and grease. These pretreatment structures may be as simple as a buffer zone or may be something more complex, such as an oil and grease separator. The types of infiltration structures discussed include infiltration trenches, porous pavements, concrete grids, and modular pavements.

9.4.1 Vegetated Filter Strips

Vegetated filter strips are gently sloping areas of natural vegetation or are graded and artificially planted areas used to provide infiltration, remove sediments and other pollutants, and reduce the flow and velocity of the storm water moving across the terrain. Vegetated filter strips function similarly to vegetated or grassed swales. The filter strips, however, are fairly level and treat sheetflow, whereas grassed swales are indentations (see section on Grassed Swales) and treat concentrated flows. Vegetated filter strips provide permanent storm water control measures on a site.

Vegetated filter strips are suited for areas where the soils are well drained or moderately well drained and where the bedrock and the water table are well below the surface. Vegetated filter strips will not function well on steep slopes, in hilly areas, or in highly paved areas because of the high velocity of runoff. Sites with slopes of 15% or more may not be suitable for filtering storm water flows. However, they should still be vegetated. This practice can be put into place at any time, provided that climatic conditions allow for planting.

At a minimum, a filter strip must be approximately 20 feet wide to function well. The length of the strip should be approximately 50 to 75 feet. Where slopes become steeper, the length of the strip must be increased. Forested strips are always preferred to vegetated strips, and existing vegetation is preferred to planted vegetation. In planning for vegetated strips, consider climatic conditions, since vegetation may not take hold in especially dry and/or cold regions.

Regular inspections are necessary to ensure the proper functioning of the filter strips. Removing sediments and replanting may be necessary on a regular basis. The entire area should be examined for damage due to equipment and vehicles. Vegetation should be dense. Also, the portions of the strip where erosion may have created ponding of runoff should be inspected. This situation can be eliminated by grading.

Advantages of Vegetated Filter Strips:

* Provide low to moderate treatment of pollutants in storm water while providing a natural look to a site.

* Can provide habitat for wildlife.

* Can screen noise and views if trees or high shrubs are planted on the filter strips.

* Are easily constructed and implemented.

* Are inexpensive.

Disadvantages of Vegetated Filter Strips:

* Are not effective for high-velocity flows (large paved areas or steep slopes).

* Require significant land space.

* May have a short useful life due to clogging by sediments and oil and grease.

9.4.2 Grassed Swales

Grassed swales are vegetated depressions used to transport, filter, and remove sediments. Grassed swales control high runoff rates by reducing the speed of the runoff and by reducing the volume of the runoff through infiltration of the storm water. Pollutants are removed because runoff travels slowly and infiltrates into the soil and because the vegetation in the grassed swale works as a filter or strainer.

Grassed swales are suitable for most areas where storm water runoff is low. Certain factors will affect the operation of grassed swales, including soil type, land features, and the depth of the soil from the surface to the water table (i.e., the top of the drenched portion of the soil or bedrock layer). The soil must be permeable for runoff to be able to infiltrate well. Sandy soils will not hold vegetation well nor form a stable channel structure. Steep slopes will increase runoff rates and create greater potential for erosion. Storm water flows will not be easily absorbed where the water table is near the surface. Swales are most useful for sites smaller than 10 acres. Even without highly permeable soils, swales reduce velocity and thus are useful.

Grassed swales usually do not work well for construction runoff because the runoff has high sediment loads.

The channel of the swale should be as level as possible to maximize infiltration. Side slopes in the swale should be designed to no steeper than 3:1 to minimize channel erosion. Plans should consider (1) the use of existing topography and existing drainage patterns and (2) the highest flow rate that is expected from a typical storm to determine the most practical size for the swale (in keeping with State or local requirements).

The swale should be tilled before grass is planted, and a dense cover of grasses should be planted in the swale. The location of the swale will determine the best type of vegetation (e.g., if the swale runs next to a road, then the grass chosen should be resistant to the use of deicing salts in northern states).

Check dams (i.e., earthen or log structures) may be installed in the swales to reduce runoff speed and increase infiltration. Planners should also consider the design of the outlet at the end of the swale so that the runoff is released from the swale at a low rate (see section on Outlet Protection).

Maintenance activities for the swales include those practices needed to maintain healthy, dense vegetation and to retain efficient infiltration and movement of the storm water into and through the swale. Periodic mowing, reseeding, and weed control are required to maintain pollutant removal efficiency. The swale and channel outlet should be kept free from sediment buildup, litter, brush, or fallen tree limbs.

Periodic inspections will identify erosion problems or damaged areas. Damaged or eroded areas of the channel should be repaired. Areas with damaged vegetation should be reseeded immediately.

Advantages of Grassed Swales:

* Are easily designed and constructed.

* Provide moderate removal of sediments if properly constructed and maintained.

* May provide a wildlife habitat.

* Are inexpensive.

* Can replace curb and gutter systems.

* Can last for long periods of time if well maintained.

Disadvantages of Grassed Swales:

* Cannot control runoff from very large storms.

* If they do not drain properly between storms, can encourage nuisance problems such as mosquitos, ragweed, dumping, and erosion.

* Are not capable of removing significant amounts of soluble nutrients.

* Cannot treat runoff with high sediment loadings.

9.4.3 Constructed Wetlands

In more recent years, constructed wetlands have been used as a BMP to treat wastewater effluent. Most wetland processes involve the growth of rooted emergent plants, such as reeds and bulrushes, in an artificial bed and the passage of wastewater either across the surface of the wetland (surface-flow systems), or through the growing medium in which the wetland plants are rooted (subsurface-flow or root zone systems).

Surface-Flow Wetlands: The surface-flow wetland approach utilizes the stems of wetland plants as the main site for effluent treatment. In this method, beds of emergent wetland plants, such as reeds or bulrushes, are flooded with pretreated effluent which is retained within the wetland system for a predetermined period prior to discharge.

Rooted aquatic plants are planted in a soil substrate within a constructed earthen basin. The basin may be lined or not, depending on natural soil permeability and groundwater protection requirements. The systems are generally designed to allow wastewater effluent following pretreatment to flow at a depth of 1 to 2 inches up to 12 to 18 inches through the basin in a plug flow pattern. The Water Environment Federation's design manual, Natural Systems for Wastewater Treatment, designates these systems as Free Water Surface systems, referring to the nature of the surface flow.

Surface-flow wetland plant stems provide a substratum for the microorganisms which achieve the desired effluent treatment. Wetland processes result in an accumulation of organic material in the bottom of the system where microorganisms also occur in high densities and further enhance effluent treatment, particularly in terms of nitrogen elimination and anaerobic decomposition of detrital material to carbon dioxide and organic acids.

The City of Arcata, California, has been experimenting with twelve 6- x 10-meter pilot project cells, two 2½-acre wetland treatment cells, and 31 acres of effluent-receiving flow wetlands.[116]

---

[116] See Robert A. Gearheart, "Use of Constructed Wetlands to Treat Domestic Wastewater, City of Arcata, California," in Proceedings of the

Subsurface-Flow Wetlands: The subsurface-flow wetlands system is similar to the surface-flow systems except that the basin is filled with aggregate, such as gravel or crushed stone, to a depth of 12 to 24 inches, in which the aquatic vegetation is planted and through which the wastewater flows with no visible surface flow. Wastewater treatment occurs in the growing medium, principally as a consequence of the growth of wetland plant rhizomes, which are claimed to enhance the hydraulic conductivity of the growth medium and introduce oxygen into adjacent areas of the growing medium. These systems have likewise been designed as Vegetated Submerged Bed systems by the Water Environment Federation's design manual.

In part due to the warmer climate and availability of land for siting, these constructed wetlands systems have more commonly been used in southern regions of the U.S. Of the 154 constructed wetlands systems (operational, constructed, and planned) inventoried in the U.S. as of March 1990 (by Sherwood Reed under contract with the EPA), 97 were located in EPA Regions 4 and 6, the southeast and southwest.

In spite of the number of these systems in use, the shortage of meaningful data and lack of understanding regarding the physical and biochemical processes taking place have resulted in slow progress toward a sound design approach. This problem has led to unexpected difficulties with a number of the constructed wetlands systems in operation. Both gravel substrate, subsurface-flow systems, and soil substrate, surface-flow systems have encountered serious problems, in some cases jeopardizing the continued use of the constructed wetlands technology in those locations.

Still, the ongoing popularity of the constructed wetlands technology primarily derives from the twofold promise of lower costs and little requirement for operation and maintenance compared to conventional technology. The EPA construction grants program encouraged the use of the technology due to the Innovative/Alternative grant bonus received by grantees selecting constructed wetlands systems of either type. The

Tennessee Valley Authority has also played a key role in encouraging the use of these systems, especially in the Tennessee Valley area.[117]

9.4.4 Level Spreaders

Level spreaders are devices used at storm water outlets to spread out collected storm water flows into sheetflow (runoff that flows over ground surface in a thin, even layer). Typically, a level spreader consists of a depression in the soil surface that spreads the flow onto a flat area across a gentle slope. Level spreaders then release the storm water flow onto level areas stabilized by vegetation to reduce speed and increase infiltration.

Level spreaders are most often used as an outlet for temporary or permanent storm water conveyances or dikes. Runoff that contains high sediment loads should be treated in a sediment-trapping device prior to release into a level spreader.

The length of the spreader depends upon the amount of water that flows through the conveyance. Larger volumes of water need more space to even out. Level spreaders are generally used with filter strips (see section on Vegetated Filter Strips). The depressions are seeded with vegetation (see section on Permanent Seeding).

Level spreaders should not be used on soil that might erode easily. They should be constructed on natural soils and not on fill material. The entrance to the spreader should be level so that the flow can spread out evenly.

The spreader should be inspected after every large storm event to check for damage. Heavy equipment and other traffic should be kept off the level spreader because these vehicles may compact the soil or disturb the grade of the slope. If ponding or erosion channels develop, the spreader should be regraded. Dense vegetation should be maintained and damaged areas reseeded as needed.

---

[117] See Robert J. Freeman, Jr., "Constructed Wetlands Experience in the Southeast," in Proceedings of the U.S. EPA Municipal Wastewater Treatment Technology Forum - 1991 (U.S. EPA, Sept. 1991).

Advantages of Level Spreaders:

* Reduce storm water flow velocity, encourage sedimentation and infiltration.

* Are relatively inexpensive to install.

Disadvantages of Level Spreaders:

* Can easily develop "short circuiting" (concentration of flows into small streams instead of sheetflow over the spreader) because of erosion or other disturbance.

* Cannot handle large quantities of sediment-laden storm water.

9.4.5 Infiltration Trenches

An infiltration trench usually consists of a long, narrow excavation ranging from 3 to 12 feet deep. The trench is filled with stone, which allows for temporary storage of storm water runoff in the open spaces between the stones. The stored storm water infiltrates into the surrounding soil or drains into underground pipes through holes and is then routed to an outflow point. Infiltration trenches are designed to remove both fine sediments and soluble pollutants rather than larger, coarse pollutants.

Infiltration trenches should be restricted to areas with certain soil, groundwater, slope, area, and pollutant conditions. For example, infiltration trenches will not operate well in soils that have high clay contents, silt/clay soils, sandy/clay loams, or soils that have been compacted. Trenches should not be sited over fill soils because such soils are unstable. Hardened soils are often not suitable for infiltration trenches because these types of soils do not easily absorb water. Infiltration practices in general should not be used to manage contaminated storm water.

The drainage area contributing runoff to a single trench should not exceed 5 acres. Construction of trenches should not start until after all land-disturbing activities have ceased so that runoff with high levels of sediment does not fill in the structure.

If slopes draining into the trench are steeper than 5%, the runoff will enter the trench too fast and will overwhelm the infiltration capacity of the soil, causing overflow. The depth from the bottom of the trench to the bedrock layer and the seasonal high water table must be at least 3 feet. Infiltration trenches may not be suitable in areas where there are cold winters and deep frost levels.

Pretreatment of runoff before it is channeled to the trench is important to efficient operation because pretreatment removes sediment, grit, and oil. Reducing the pollutant load in the runoff entering the trench lengthens trench life. One method of pretreatment is to install a buffer zone just above the trench to act as a filter (see section on Buffer Zones). In addition, a layer of filter fabric 1 foot below the bottom of the trench can be used to trap the sediments that get through the buffer strip. If excavation around the trenches is necessary, the use of light-duty equipment will avoid compacting, which could cause a loss of infiltration capability.

Infiltration trenches should be inspected at least once per year and after major rainfall events. Debris should be removed from all areas of the trench, especially the inlets and overflow channels. Dense vegetative growth should be maintained in buffer areas surrounding the trench.

Test wells can be installed in every trench to monitor draining times and provide information on how well the system is operating. Daily test well monitoring is necessary, especially after large storm events. If the trench does not drain after 3 days, it usually means that the trench is clogged.

Advantages of Infiltration Trenches:

* Preserve the natural water balance of the site.

* Are effective for small sites.

* Remove pollutants effectively.

Disadvantages of Infiltration Trenches:

* Require high maintenance when sediment loads are heavy.

* Have short life span, especially if not maintained properly.

* May be expensive (cost of excavation and fill material).

9.4.6 Porous Pavement/Concrete Grids and Modular Pavements

Porous pavement, concrete grids, and modular pavements allow storm water to infiltrate so that the speed and amount of runoff from a site can be reduced.

Porous Pavement: Can be either asphalt or concrete. With porous asphalt pavement, runoff infiltrates through a porous asphalt layer into a stone "reservoir" layer. Storm water runoff filters through the stone reservoir into the underlying subsoil or drains into underground pipes through holes and is routed away. The bottom and sides of the stone reservoir are lined with filter fabric to prevent the movement of soils into the reservoir area.

Porous Concrete Pavement: Is made out of a special concrete mix that has a high number of open spaces between the particles and a coarse surface texture. These open spaces allow runoff to pass through the surface to lower levels. This type of pavement can be placed directly on graded soils. When a subbase is used for stability, 6 inches of sand is placed under the concrete mixture. Up to 6 inches of storm water can be held on the surface of the pavement and within the concrete.

Concrete Grids and Modular Pavement: These are made out of precast concrete, poured-in-place concrete, brick, or granite. These types of pavements can also reduce the loading and concentration of pollutants in the runoff.

> Concrete grids and modular pavements are designed and/or constructed so that they have open spaces within the pavement through which storm water can infiltrate into the ground. These open spaces can be filled with gravel or sand or have vegetation growing out of them.

These structures are usually only suitable for low-volume parking areas (1/4 acre to 10 acres) and lightly used access roads. However, areas that are expected to get moderate or high volumes of traffic or heavy equipment can use conventional pavements (for the heavy traffic areas) that are sloped to drain to areas with the porous pavements. These pavements are not effective in drainage areas that receive runoff containing high levels of sediment.

The soil types over which concrete grids and modular pavement are to be placed should allow for rapid drainage through the pores in the pavement. These pavements are not recommended for sites with slopes steeper than 5% or sites with high water tables, shallow bedrock, fill soils, or localized clay lenses, which are conditions that would limit the ability of the runoff to infiltrate into surface scils. For example, the water table and bedrock should be at least 3 feet below the bottom of the stone reservoir. Porous pavement will not operate well in windy areas where sediment will be deposited on the porous pavement.

Construction of these pavements should be timed so that installation occurs on the site after other construction activities are finished and the site has been stabilized. Therefore, sediments are less likely to be tracked or carried on to the surface.

Proper installation of these pavements requires a high level of construction expertise and workmanship. Only contractors who are familiar with the installation of these pavements should be used.

Designers of porous pavement areas should consider sediment and erosion control. Sediments must be kept away from the pavement area because they can clog the pores. Controls to consider for sediments include a diversion berm (i.e., earthen mound) around the edge of the pavement area to block the flow of runoff from certain drainages onto the pavement, or other filtering controls such as silt fences. Deicing salt mixtures, sands, or ash also may clog pores and should not be used for snow removal. Signs should be posted to prohibit these activities.

Since the infiltration of storm water runoff may contaminate groundwater sources, these pavements are not suitable for areas close to drinking water wells (at least 100 feet away is recommended).

Maintenance of the surface is very important. For porous pavements, this includes vacuum sweeping at least four times per year followed by high-pressure hosing to reduce the chance of sediments clogging the pores of the top layer. Potholes and cracks can be filled with typical patching mixes unless more than 10% of the surface area needs repair. Spot clogging may be fixed by drilling half-inch holes through the porous pavement layer every few feet.

The pavement should be inspected several times the first few months after installation and then annually. Inspections after large storms are necessary to check for pools of water. These pools may indicate clogging. The condition of adjacent vegetated filter strips, silt fences, or diversion dikes should also be inspected.

Concrete grids and modular pavements should be designed in accordance with manufacturers' recommendations. Designers also need information on soils, depth to the water table, and storm water runoff quantity and quality.

Maintenance of concrete grids and modular pavements is similar to that of the porous pavements; however, turf maintenance such as mowing, fertilizing, and irrigation may be needed where vegetation is planted in the open spaces.

Advantages of Porous Pavements, Concrete Grids, and Modular Pavements:

* Provide erosion control by reducing the speed and quantity of the storm water runoff from the site.

* Provide some treatment to the water by removing pollutants.

* Reduce the need for curbing and storm sewer installation and expansion.

* Improve road safety by providing a rougher surface.

* Provide some recharge to local aquifers.

* Are cost effective because they take the place of more expensive and complex treatment systems.

Disadvantages of Porous Pavements, Concrete Grids, and Modular Pavements:

* Can be more expensive than typical pavements.

* Are easily clogged with sediment and/or oil; however, pretreatment and proper maintenance will prevent this problem.

* May cause groundwater contamination.

* Are not structurally suited for high-density traffic or heavy equipment.

* Asphalt pavements may break down if gasoline is spilled on the surface.

* Are less effective when the subsurface is frozen.

Appendix A

GLOSSARY OF TERMS

Aeration: A process which promotes biological degradation of organic matter. The process may be passive (as when waste is exposed to air) or active (as when a mixing or bubbling device introduces the air).

Aliquot: A discrete sample used for analysis.

Backfill: Earth used to fill a trench or an excavation.

Baffles: Fin-like devices installed vertically on the inside walls of liquid waste transport vehicles that are used to reduce the movement of the waste inside the tank.

Berm: An earthen mound used to direct the flow of runoff around or through a structure.

Best Management Practice (BMP): Schedules of activities, prohibitions of practices, maintenance procedures, and other management practices to prevent or reduce the pollution of waters of the U.S. BMPs also include treatment requirements, operating procedures, and practices to control facility site runoff, spillage or leaks, sludge or waste disposal, or drainage from raw material storage.

Biochemical Oxygen Demand (BOD): The quantity of oxygen consumed during the biochemical oxidation of matter over a specified period of time, usually 5 days (BOD_5).

Biodegradable: The ability to break down or decompose under natural conditions and processes.

Boom: 1. A floating device used to contain oil on a body of water. 2. A piece of equipment used to apply pesticides from ground equipment such as a tractor or truck.

Buffer Strip or Zone: Strips of grass or other erosion-resistant vegetation between a waterway and an area of more intensive land use.

By-product: Material, other than the principal product, that is generated as a consequence of an industrial process.

Calibration: A check of the precision and accuracy of measuring equipment.

CERCLA: Comprehensive Environmental Response, Compensation, and Liability Act, 42 U.S.C. 9601 *et seq.*

Chain-of-Custody: Procedures used to minimize the possibility of tampering with samples.

Chemical Oxygen Demand (COD): Measurement of all the oxidizable matter found in a runoff sample, a portion of which could deplete dissolved oxygen in receiving waters.

Chock: A block or wedge used to keep rolling vehicles in place.

Clay Lens: A naturally occurring, localized area of clay that acts as an impermeable layer to runoff infiltration.

Composite Sample: Used to determine "average" loadings or concentrations of pollutants, such samples are collected at regular time intervals and pooled into one large sample; can be developed on time or flow rate.

Concrete Aprons: A pad of nonerosive material designed to prevent scour holes developing at the outlet ends of culverts, outlet pipes, grade-stabilization structures, and other water control devices.

Conduit: Any channel or pipe for transporting the flow of water.

Conveyance: Any natural or manmade channel or pipe in which concentrated water flows.

Corrosion: The dissolving and wearing away of metal caused by a chemical reaction such as between water and the pipes that the water contacts, chemicals touching a metal surface, or contact between two metals.

Culvert: A covered channel or a large-diameter pipe that directs water flow below the ground level.

CWA: Clean Water Act (formerly referred to as the Federal Water Pollution Control Act), 33 U.S.C. 1251 *et seq.*

Denuded: Land stripped of vegetation such as grass, or land that has had vegetation worn down due to impacts from the elements or humans.

Detention Ponds: A surface water impoundment constructed to hold and manage storm water runoff.

Dike: An embankment to confine or control water, often built along the banks of a river to prevent overflow of lowlands; a levee.

Director: The EPA Regional Administrator or an authorized representative.

Discharge: A release or flow of storm water or other substance from a conveyance or storage container.

Drip Guard: A device used to prevent drips of fuel or corrosive or reactive chemicals from contacting other materials or areas.

Effluent: Any discharge flowing from a conveyance.

Emission: Pollution discharged into the atmosphere from smokestacks, other vents, and surface areas of commercial or industrial facilities and from motor vehicle, locomotive, or aircraft exhausts.

EPCRA: Emergency Planning and Community Right-to-Know Act, 42 U.S.C. 11001-11050. (also known as SARA Title III). The primary goal of EPCRA is to facilitate public awareness and emergency response planning for chemical hazards. To fulfill this purpose, EPCRA requires that certain companies that manufacture, process, and use chemicals in specified quantities must file written reports, provide notification of spills/releases, and maintain toxic chemical inventories.

Erosion: The wearing away of land surface by wind or water. Erosion occurs naturally from weather or runoff, but can be intensified by land-clearing practices related to farming, residential or industrial development, road building, or timber cutting.

Excavation: The process of removing earth, stone, or other materials.

Fertilizer: Materials such as nitrogen and phosphorus that provide nutrients for plants. Commercially sold fertilizers may contain other chemicals or may be in the form of processed sewage sludge.

Filter Fabric: Textile of relatively small mesh or pore size that is used to (a) allow water to pass through while keeping sediment out (permeable), or (b) prevent both runoff and sediment from passing through (impermeable).

Filter Strip: Usually long, relatively narrow area of undisturbed or planted vegetation used to retard or collect sediment for the protection of watercourses, reservoirs, or adjacent properties.

First Flush: Individual sample taken during the first 30 minutes of a storm event. The pollutants in this sample can often be used as a screen for nonstorm water discharges, since such pollutants are flushed out of the system during the initial portion of the discharge.

Flange: A rim extending from the end of a pipe; can be used as a connection to another pipe.

Flow Channel Liner: A covering or coating used on the inside surface of a flow channel to prevent the infiltration of water to the ground.

Flowmeter: A gauge that shows the speed of water moving through a conveyance.

Flow-Proportional Composite Sample: Combines discrete aliquots of a sample collected over time, based on the flow of the waste stream being sampled. There are two methods used to collect this type of sample. One collects a constant sample volume at time intervals which vary based on stream flow. The other collects aliquots at varying volumes based on stream flow, at constant time intervals.

Flow-Weighted Composite Sample: Means a composite sample consisting of a mixture of aliquots collected at a constant time interval, where the volume of each aliquot is proportional to the flow rate of the discharge.

General Permit: A permit issued under the NPDES program to cover a certain class or category of storm water discharges. These permits allow for a reduction in the administrative burden associated with permitting storm water discharges associated with industrial activities. See discussion in Chapters 2 and 3 for detailed guidance on General Permits.

Grab Sample: A discrete sample which is taken from a waste stream on a one-time basis with no regard to flow or time; instantaneous sample that is analyzed separately.

Grading: The cutting and/or filling of the land surface to a desired slope or elevation.

Hazardous Substance: 1. Any substance that poses a threat to human health and/or the environment. Hazardous substances can be toxic, corrosive, ignitable, explosive, or chemically reactive. 2. Any substance required by EPA to be reported if a designated quantity of the substance is spilled in the waters of the U.S. or if otherwise emitted into the environment.

Hazardous Substance (CERCLA Definition): Under CERCLA 101(14), any element, compound, mixture, solution, or substance which, when released to the environment, may present substantial danger to public health/welfare or the environment. Also includes (a) any substance designated pursuant to section 311(b)(2)(A) of the Clean Water Act; (b) any element, compound, mixture, solution, or substance designated pursuant to section 102 of CERCLA; (c) any hazardous waste having the characteristics identified under or listed pursuant to section 3001 of the Resource Conservation and Recovery Act; (d) any toxic pollutant listed under section 307(a) of the Clean Water Act; (e) any hazardous air pollutant listed under section 112 of the Clean Air Act; and (f) any imminently hazardous chemical substance or mixture so identified pursuant to the Toxic Substances Control Act. The CERCLA definition excludes petroleum (including crude oil not otherwise specifically listed or designated as a hazardous substance under any of the above laws), natural gas, natural gas liquids, liquefied natural gas, or synthetic gas usable for fuel (or mixtures of natural gas and such synthetic gas). The definition of hazardous substances in CERCLA is broader than the definition of hazardous wastes under RCRA.

Hazardous Waste: By-products of human activities that can pose a substantial or potential hazard to human health or the environment when improperly managed. Possesses at least one of four characteristics (ignitability, corrosivity, reactivity, or toxicity), or appears on special EPA lists.

Hazardous Waste (RCRA Definition): A solid waste which because of its quantity, concentration, or physical, chemical, or infectious characteristics may (a) cause or contribute to an increase in mortality or an increase in serious irreversible or incapacitating reversible illness; or (b) pose a substantial present or potential hazard to human health or the environment when improperly treated, stored, transported, or disposed of, or otherwise managed. EPA hazardous waste regulations (40 CFR 261) specify that if a material qualifies as a solid waste, and does not qualify for an exemption, it is a hazardous waste if it is listed by 40 CFR Part 261, Subpart D, or if it exhibits any of the four hazardous waste characteristics (ignitability, reactivity, corrosivity and toxicity).

Hazardous Waste Constituent: A constituent that caused the waste to be listed as a RCRA hazardous waste under 40 CFR Part 261 Subpart D.

Holding Pond: A pond or reservoir, usually made of earth, built to store polluted runoff for a limited time.

HSWA: Hazardous and Solid Waste Amendments of 1984, which amended RCRA. The HSWA greatly increased the complexity of the RCRA regulatory program by imposing restrictions on land disposal of hazardous wastes, authorizing EPA to require corrective action for releases from hazardous waste management facilities, and instituting requirements for underground storage tanks containing petroleum and hazardous substances.

Illicit Connection: Any discharge to a municipal separate storm sewer that is not composed entirely of storm water except discharges authorized by an NPDES permit (other than the NPDES permit for discharges from the municipal separate storm sewer) and discharges resulting from fire-fighting activities.

Infiltration: 1. The penetration of water through the ground surface into subsurface soil or the penetration of water from the soil into sewer or other pipes through defective joints, connections, or manhole walls. 2. A land application technique where large volumes of wastewater are applied to land, allowed to penetrate the surface, and percolate through the underlying soil.

Inlet: An entrance into a ditch, storm sewer, or other waterway.

Intermediates: A chemical compound formed during the making of a product.

Irrigation: Human application of water to agricultural or recreational land for watering purposes.

Jute: A plant fiber used to make rope, mulch, netting, or matting.

Lagoon: A shallow pond where sunlight, bacterial action, and oxygen work to purify wastewater.

Land Application Units: An area where wastes are applied onto or incorporated into the soil surface (excluding manure-spreading operations) for treatment or disposal.

Land Treatment Units: An area of land where materials are temporarily located to receive treatment. Examples include sludge lagoons and stabilization ponds.

Landfills: An area of land or an excavation in which wastes are placed for permanent disposal, and which is not a land application unit, surface impoundment, injection well, or waste pile.

Large and Medium Municipal Separate Storm Sewer System: All municipal separate storm sewers that are either: (a) located in an incorporated place (city) with a population of 100,000 or more as determined by the latest Decennial Census by the Bureau of Census (these cities are listed in Appendices F and G of 40 CFR Part 122); or (b) located in the counties with unincorporated urbanized populations of 100,000 or more, except municipal separate storm sewers that are located in the incorporated places, townships, or towns within such counties (these counties are listed in Appendices H and I of 40 CFR Part 122); or (c) owned or operated by a municipality other than those described in (a) or (b) and that are designated by the Director as part of the large or medium municipal separate storm sewer system.

Leaching: The process by which soluble constituents are dissolved in a solvent such as water and carried down through the soil.

Level Spreader: A device used to spread out storm water runoff uniformly over the ground surface as sheet flow (i.e., not through channels). The purpose of level spreaders is to prevent concentrated, erosive flows from occurring and to enhance infiltration.

Liming: Treating soil with lime to neutralize acidity levels.

Liner: 1. A relatively impermeable barrier designed to prevent leachate from leaking from a landfill. Liner materials include plastic and dense clay. 2. An insert or sleeve for sewer pipes to prevent leakage or infiltration.

Liquid Level Detector: A device that provides continuous measures of liquid levels in liquid storage areas or containers to prevent overflows.

Materials Management Practices: Practices used to limit the contact between significant materials and precipitation. These may include structural or nonstructural controls such as dikes, berms, sedimentation ponds, vegetation strips, spill response plans, and so on.

Material Storage Areas: On-site locations where raw materials, products, final products, byproducts, or waste materials are stored.

Mulch: A natural or artificial layer of plant residue or other materials covering the land surface which conserves moisture, holds soil in place, aids in establishing plant cover, and minimizes temperature fluctuations.

Noncontact Cooling Water: Water used to cool machinery or other materials without directly contacting process chemicals or materials.

Notice of Intent (NOI): An application to notify the permitting authority of a facility's intention to be covered by a general permit; exempts a facility from having to submit an individual or group application.

NPDES: The National Pollutant Discharge Elimination System used by EPA to control the discharge of pollutants to waters of the U.S. See 40 CFR 122.2 for further guidance.

NPDES Permit: An authorization, license, or equivalent control document issued by EPA or an approved State agency to implement the requirements of the NPDES program.

Oil and Grease Traps: Devices which collect oil and grease, removing them from water flows.

Oil Sheen: A thin, glistening layer of oil on water.

Oil/Water Separator: A device installed, usually at the entrance to a drain, which removes oil and grease from water flows entering the drain.

Organic Pollutants: Substances containing carbon which may cause pollution problems in receiving streams.

Organic Solvents: Liquid organic compounds capable of dissolving solids, gases, or liquids.

Outfall: The point, location, or structure where wastewater or drainage discharges from a sewer pipe, ditch, or other conveyance to a receiving body of water.

Permeability: The quality of a soil that enables water or air to move through it. Usually expressed in inches/hour or inches/day.

Permit: An authorization, license, or equivalent control document issued by EPA or an approved State agency to implement the requirements of an environmental regulation; e.g., a permit to operate a wastewater treatment plant or to operate a facility that may generate harmful emissions.

Permit Issuing Authority (or Permitting Authority): The State agency or EPA Regional office which issues environmental permits to regulated facilities.

Plunge Pool: A basin used to slow flowing water, usually constructed to a design depth and shape. The pool may be protected from erosion by various lining materials.

Pneumatic Transfer: A system of hoses which uses the force of air or other gas to push material through; used to transfer solid or liquid materials from tank to tank.

Point Source: Any discernible, confined, and discrete conveyance, including but not limited to any pipe, ditch, channel, tunnel, conduit, well, discrete fissure, container, rolling stock, concentrated animal feeding operation, or vessel or other floating craft, from which pollutants are or may be discharged. This term does not include return flows from irrigated agriculture or agricultural storm water runoff.

Pollutant: Any dredged spoil, solid waste, incinerator residue, filter backwash, sewage, garbage, sewage sludge, munitions, chemical wastes, biological materials, radioactive materials (except those regulated under the Atomic Energy Act of 1954, as amended [42 U.S.C. 2011 *et seq.*]), heat, wrecked or discharged equipment, rock, sand, cellar dirt, and industrial, municipal, and agricultural waste discharged into water. It does not mean: (a) sewage from vessels; or (b) water, gas, or other material which is injected into a well to facilitate production of oil or gas, or water derived in association with oil and gas production and disposed of in a well, if the well used either to facilitate production or for disposal purposes is approved by the authority of the State in which the well is located, and if the State determines that the injection or disposal will not result in the degradation of ground or surface water resources. (See CWA Section 502(6)).

Porous Pavement: A human-made surface that will allow water to penetrate through and percolate into soil (as in porous asphalt pavement or concrete). Porous asphalt pavement is comprised of irregular shaped crush rock precoated with asphalt binder. Water seeps through into lower layers of gravel for temporary storage, then filters naturally into the soil.

Precipitation: Any form of rain or snow.

Preventative Maintenance Program: A schedule of inspections and testing at regular intervals intended to prevent equipment failures and deterioration.

Process Wastewater: Water that comes into direct contact with or results from the production or use of any raw material, intermediate product, finished product, by-product, waste product, or wastewater.

PVC (Polyvinyl Chloride): A plastic used in pipes because of its strength; does not dissolve in most organic solvents.

Raw Material: Any product or material that is converted into another material by processing or manufacturing.

RCRA: Resource Conservation and Recovery Act, 42 U.S.C. 6901 *et seq.*

Recycle: The process of minimizing the generation of waste by recovering usable products that might otherwise become waste. Examples are the recycling of aluminum cans, wastepaper, and bottles.

Reportable Quantity (RQ): The quantity of a hazardous substance or oil that triggers reporting requirements under CERCLA, the Clean Water Act, or other environmental laws. If a substance is released in amounts exceeding its RQ, the release must be reported to the National Response Center, the State Emergency Response Commission, and/or other federal, state, and local authorities. (See Dennison, *Environmental Reporting, Recordkeeping, and Inspections: Compliance Guide for Business* *and Industry* (Van Nostrand Reinhold, 1995) for additional guidance on environmental reporting requirements).

Residual: Amount of pollutant remaining in the environment after a natural or technological process has taken place, e.g., the sludge remaining after initial wastewater treatment, or particulates remaining in air after the air passes through a scrubbing or other pollutant removal process.

Retention: The holding of runoff in a basin without release except by means of evaporation, infiltration, or emergency bypass.

Retrofit: The modification of storm water management systems in developed areas through the construction of wet ponds, infiltration systems, wetland plantings, stream bank stabilization, and other BMP techniques for improving water quality. A retrofit can consist of the construction of a new BMP in the developed area, the enhancement of an older storm water management structure, or a combination of improvement and new construction.

Reverse Meniscus: The curved upper surface of a liquid in a container.

Rill Erosion: The formation of numerous, closely spread streamlets due to uneven removal of surface soils by storm water or other water.

Riparian Habitat: Areas adjacent to rivers and streams that have a high density, diversity, and productivity of plant and animal species relative to nearby uplands.

Runon: Storm water surface flow or other surface flow which enters property other than that where it originated.

Runoff: That part of precipitation, snow melt, or irrigation water that runs off the land into streams or other surface water. It can carry pollutants from the air and land into the receiving waters.

Runoff Coefficient: Means the fraction of total rainfall that will appear at the conveyance as runoff.

Sanitary Sewer: A system of underground pipes that carries sanitary waste or process wastewater to a treatment plant.

Sanitary Waste: Domestic sewage.

SARA: Superfund Amendments and Reauthorization Act of 1986, which amended CERCLA. SARA Title III is commonly known as the Emergency Planning and Community Right-to-Know Act (EPCRA).

Scour: The clearing and digging action of flowing water, especially the downward erosion caused by stream water in sweeping away mud and silt from the stream bed and outside bank of a curved channel.

Sealed Gate: A device used to control the flow of liquid materials through a valve.

Secondary Containment: Structures, usually dikes or berms, surrounding tanks or other storage containers and designed to catch spilled material from the storage containers.

Section 313 Water Priority Chemical: A chemical or chemical categories which are: (a) are listed at 40 CFR 372.65 pursuant to Section 313 of the Emergency Planning and Community Right-to-Know Act (EPCRA); (b) are present at or above threshold levels at a facility subject to EPCRA Section 313 reporting requirements; and (c) that meet at least one of the following criteria: (i) are listed in Appendix D of 40 CFR Part 122 on either Table II (organic priority pollutants), Table III (certain metals, cyanides, and phenols), or Table V (certain toxic pollutants and hazardous substances); (ii) are listed as a hazardous substance pursuant to Section 311 (b)(2)(A) of the CWA at 40 CFR 116.4; or (iii) are pollutants for which EPA has published acute or chronic water quality criteria.

Sediment Trap: A device for removing sediment from water flows; usually installed at outfall points.

Sedimentation: The process of depositing soil particles, clays, sands, or other sediments that were picked up by flowing water.

Sediments: Soil, sand, and minerals washed from land into water, usually after rain. They pile up in reservoirs, rivers, and harbors, destroying fish-nesting areas and holes of water animals and cloud the water so that needed sunlight might not reach aquatic plants. Careless farming, mining, and building activities will expose sediment materials, allowing them to be washed off the land after rainfalls.

Sheet Erosion: Erosion of thin layers of surface materials by continuous sheets of running water.

Sheetflow: Runoff which flows over the ground surface as a thin, even layer, not concentrated in a channel.

Shelf Life: The time for which chemicals and other materials can be stored before becoming unusable due to age or deterioration.

Significant Materials: Include, but are not limited to: raw materials; fuels; materials such as solvents, detergents, and plastic pellets; finished materials such as metallic products; raw materials used in food processing or production; hazardous substances designated under CERCLA section 101(14); any chemical the facility is required to report pursuant to EPCRA section 313; fertilizers; pesticides; and waste products such as ashes, slag, and sludge that have a potential to be released with storm water discharges. (See 40 CFR 122.26(b)(12)).

Significant Spills: Includes, but is not limited to: releases of oil or hazardous substances in excess of reportable quantities under CWA Section 311. (See 40 CFR 110.10 and CFR 117.21 or 40 CFR 302.4).

Slag: Nonmetal-containing waste left over from the smelting and refining of metals.

Slide Gate: A device used to control the flow of water through storm water conveyances.

Sloughing: The movement of unstabilized soil layers down a slope due to excess water in the soils.

Sludge: A semisolid residue from any of a number of air or water treatment processes. Sludge can be a hazardous waste.

Soil: The unconsolidated mineral and organic material on the immediate surface of the earth that serves as a natural medium for the growth of plants.

Solids Dewatering: A process for removing excess water from solids to lessen the overall weight of the wastes.

Source Control: A practice or structural measure to prevent pollutants from entering storm water runoff or other environmental media.

Spent Solvent: A liquid solution that has been used and is no longer capable of dissolving solids, gases, or liquids.

Spill Guard: A device used to prevent spills of liquid materials from storage containers.

Spill Prevention Control and Countermeasures Plan (SPCC): Plan consisting of structures, such as curbing, and action plans to prevent and respond to spills of hazardous substances as defined in the Clean Water Act.

Stopcock Valve: A small valve for stopping or controlling the flow of water or other liquid through a pipe.

Storm Drain: A slotted opening leading to an underground pipe or an open ditch for carrying surface runoff.

Storm Water: Runoff from a storm event, snow melt runoff, and surface runoff and drainage.

Storm Water Discharge Associated with Industrial Activity: [See 40 CFR 122.26(b)(14)]. The discharge from any conveyance which is used for collecting and conveying storm water and which is directly related to manufacturing, processing, or raw materials storage areas at an industrial plant. The term does not include discharges from facilities or activities excluded from the NPDES program under 40 CFR Part 122. For the categories of industries identified in subparagraphs (i) through (x) of this subsection, the term includes, but is not limited to, storm water discharges from industrial plant yards; immediate access roads and rail lines used or traveled by carriers of raw materials, manufactured products, waste material, or by-products used or created by the facility; material-handling sites; refuse sites; sites used for the application or disposal of process wastewaters (as defined at 40 CFR 401); sites used for the storage and maintenance of material-handling equipment; sites used for residual treatment, storage, or disposal; shipping and receiving areas; manufacturing buildings; storage areas (including tank farms) for raw materials, and intermediate and finished products; and areas where industrial activity has taken place in the past and significant materials remain and are exposed to storm water. For the categories of industries identified in subparagraph (xi), the term includes only storm water discharges from all the areas (except access roads and rail lines) that are listed in the previous sentence where material-handling equipment or activities, raw materials, intermediate products, final products, waste material, by-products, or industrial machinery *are exposed to storm water.* For the purposes of this paragraph, material-handling activities include: the storage, loading and unloading, transportation, or conveyance of any raw material, intermediate product, finished product, by-product, or waste product. The term excludes areas located on plant lands separate from the plant's industrial activities, such as office buildings and accompanying parking lots as long as the drainage from the excluded areas is not mixed with storm water drained from the above described areas. Industrial facilities (including industrial facilities that are Federally, State, or municipally owned or operated that meet the description of the facilities listed in this paragraph (i)-(xi) include those facilities designated under the provision of 122.26(a)(1)(v). *(See Table 2.2 in Chapter 2 for a listing of the categories of facilities that are considered to be engaging in "industrial activity" (subsections (i) through (xi)).*

Subsoil: The bed or stratum of earth lying below the surface soil.

Sump: A pit or tank that catches liquid runoff for drainage or disposal.

Surface Impoundment: Treatment, storage, or disposal of liquid wastes in ponds.

Surface Water: All water naturally open to the atmosphere (rivers, lakes, reservoirs, streams, wetlands impoundments, seas, estuaries, etc.); also refers to springs, wells, or other collectors which are directly influenced by surface water.

Swale: An elongated depression in the land surface that is at least seasonally wet, is usually heavily vegetated, and is normally without flowing water. Swales direct storm water flows into primary drainage channels and allow some of the storm water to infiltrate into the ground surface.

Tarp: A sheet of waterproof canvas or other material used to cover and protect materials, equipment, or vehicles.

Time Composite Sample: Prepared by collecting fixed-volume aliquots at specified time intervals and combining into a single sample for analysis.

Topography: The physical features of a surface area including relative elevations and the position of natural and human-made features.

Toxic Pollutants: Any pollutant listed as toxic under CWA Section 501 (a)(1) or, in the case of "sludge use or disposal practices," any pollutant identified in regulations implementing CWA Section 405(d). Refer to 40 CFR Part 122, Appendix D for further guidance.

Treatment: The act of applying a procedure or chemicals to a substance to remove undesirable pollutants.

Tributary: A river or stream that flows into a larger river or stream.

Turbidity: Describes the capability of light to pass through water.

Underground Storage Tank (UST): *RCRA definition* - Any one or combination of tanks (including its connecting underground pipes) used to contain an accumulation of regulated substances, and the volume of which (including the volume of the underground pipes) is 10% or more beneath the surface of the ground. Regulated substances include hazardous chemical products regulated under CERCLA and petroleum products. Some tank uses are exempt from regulation, including septic tanks, residential/agricultural fuel or heating oil tanks, and wastewater collection systems. (See 40 CFR 280.12 for the complete RCRA definition.)

Waste Pile: Any noncontainerized accumulation of solid, nonflowing waste that is used for treatment or storage.

Water Table: The depth or level below which the ground is saturated with water.

Waters of the United States: "Waters of the United States" are defined in **33 CFR 328.3** as: "(a) All waters, which are currently used, were used in the past, or may be susceptible to use in interstate or foreign commerce, including all waters which are subject to the ebb and flow of the tide; (b) All interstate waters, including interstate "wetlands"; (c) All other waters such as intrastate lakes, rivers, streams (including intermittent streams), mud flats, sand flats, "wetlands," sloughs, prairie potholes, wet meadows, playa lakes, or natural ponds, the use, degradation, or destruction of which would affect or could affect interstate or foreign commerce ....; (d) All impoundments of waters otherwise defined as waters of the United States under this definition; (e) Tributaries of waters identified in paragraphs (a) through (d) of this definition; (f) The territorial sea; and (g) "Wetlands" adjacent to waters (other than waters that are themselves wetlands) identified in paragraphs (a) through (f) of this definition.

Waste treatment systems, including treatment ponds or lagoons designed to meet the requirements of CWA (other than cooling ponds as defined in 40 CFR 423.11 (m) which also meet the criteria of this definition) are not waters of the United States. This exclusion applies only to manmade bodies of water which neither were originally created in waters of the United States (such as disposal area in wetlands) nor resulted from the impoundment of waters of the United States."

Waterway: A channel for the passage or flow of water.

Weir: A device used to gauge the flow rate of liquid through a channel; is essentially a dam built across an open channel over which the liquid flows, usually through some type of notch.

Wet Well: A chamber used to collect water or other liquid and to which a pump is attached.

Wetlands: An area that is regularly saturated by surface or ground water and subsequently is characterized by a prevalence of vegetation that is adapted for life in saturated soil conditions. Examples include: swamps, bogs, fens, marshes, and estuaries.

Wind Break: Any device designed to block wind flow and intended for protection against any ill effects of wind.

Appendix B

FEDERAL AND STATE STORM WATER REGULATORY OFFICES

EPA REGIONAL OFFICES

USEPA Region 1

Connecticut, Massachusetts, Maine, New Hampshire, Rhode Island, Vermont

Storm Water Permits
Waste Management Division
U.S. EPA
John F. Kennedy Building
Boston, MA 02203
(617) 565-3420

USEPA Region 2

New Jersey, New York, Puerto Rico, Virgin Islands

Storm Water Permits
Waste Management Division
U.S. EPA
26 Federal Plaza
New York, NY 10278
(212) 264-2657

USEPA Region 3

Delaware, Maryland, Pennsylvania, Virginia, West Virginia, District of Columbia

Storm Water Permits
Waste Management Division
U.S. EPA
841 Chestnut Street
Philadelphia, PA 19107
(215) 597-9800

USEPA Region 4

Alabama, Florida, Georgia, Kentucky, Mississippi, North Carolina, South Carolina, Tennessee

Storm Water Permits
Waste Management Division
U.S. EPA
345 Courtland Street, N.E.
Atlanta, GA 30365
(404) 347-4727

USEPA Region 5

Illinois, Indiana, Michigan, Minnesota, Ohio, Wisconsin

Storm Water Permits
Waste Management Division
U.S. EPA
230 South Dearborn Street
Chicago, IL 60604
(312) 353-2105

USEPA Region 6

Arkansas, Louisiana, New Mexico, Oklahoma, Texas

Storm Water Permits
Waste Management Division
U.S. EPA
1445 Ross Avenue, 12th Floor
Dallas, TX 75202
(214) 655-6444

USEPA Region 7

Iowa, Kansas, Missouri, Nebraska

Storm Water Permits
Waste Management Division
U.S. EPA
726 Minnesota Avenue
Kansas City, KS 66101
(913) 551-7000

USEPA Region 8

Colorado, Montana, North Dakota, South Dakota, Utah, Wyoming

Storm Water Permits
Waste Management Division
U.S. EPA
999 18th Street, Suite 500
Denver, CO 80202
(303) 293-1603

USEPA Region 9

Arizona, California, Hawaii, Nevada, American Samoa, Guam, Trust Territories of the Pacific

Storm Water Permits
Waste Management Division
U.S. EPA
75 Hawthorne Street
San Francisco, CA 94105
(415) 744-2125

USEPA Region 10

Alaska, Idaho, Oregon, Washington

Storm Water Permits
Waste Management Division
U.S. EPA
1200 6th Avenue
Seattle, WA 98101
(206) 442-1200

STATE AGENCY CONTACTS

Alabama

Alabama Department of Environmental Management
1751 Congressman Dickinson Drive
Montgomery, AL 36130
(205) 271-7825

Alaska
(and EPA Region 10)

Alaska Department of Environmental Conservation
Division of Environmental Quality
P.O. Box #O
Juneau, AK 99811
(907) 465-2640

American Samoa

EPA Region 9

Arizona
(and EPA Region 9)

Arizona Department of Health Services
Office of Waste and Water Quality Management
2005 N. Central Avenue
Phoenix, AZ 85007
(602) 257-2305

Arkansas

Arkansas Department of Pollution Control and Ecology
NPDES Branch
8001 National Drive
Little Rock, AR 72209
(501) 562-7444

California

California Water Resources Control Board
P.O. Box 100
901 P Street
Sacramento, CA 95801
(916) 322-3132

Colorado

Colorado Department of Health
Water Quality Division
Permits and Enforcement Division
4210 E. 11th Avenue
Denver, CO 80220
(303) 331-3015

Connecticut

Department of Environmental Protection
Water Compliance and Hazardous Substances
122 Washington Street
Hartford, CT 06106
(203) 566-3245

Delaware

Department of Natural Resources and Environmental Control
Division of Water Resources
89 Kings Highway
P.O. Box 1401
Dover, DE 19903
(302) 736-4761

District of Columbia
(and EPA Region 3)

D.C. Department of Consumer and Regulatory Affairs
Environmental Control Division
5010 Overlook Avenue, S.W.
Washington, DC 20032
(202) 767-7370

Florida
(and EPA Region 4)

Department of Environmental Regulation
Division of Environmental Programs
Water Quality Planning Section
Twin Towers Office Building
2600 Blair Stone Road
Tallahassee, FL 32301
(904) 488-0780

Georgia

Georgia Department of Natural Resources
Environmental Protection Division
Water Protection Branch
Floyd Towers East
205 Butler Street, S.W.
Atlanta, GA 30334
(404) 656-4887

Guam

EPA Region 9

Hawaii

Hawaii Department of Health
P.O. Box 3378
Honolulu, HI 96801
(808) 548-6505

Idaho
(and EPA Region 10)

Idaho Department of Health and Welfare
Bureau of Water Quality
Idaho State House
450 W. State Street
Boise, ID 83720
(208) 334-4250

Illinois

Illinois Environmental Protection Agency
Division of Water Pollution Control
2200 Churchill Road
Springfield, IL 62706
(217) 782-1654

Indiana

Indiana Department of Environmental Management
105 S. Meridian Street
P.O. Box 6015
Indianapolis, IN 46225
(317) 232-8488

Iowa

Iowa Department of Natural Resources
Environmental Protection Division
Surface and Ground Water Protection Bureau
Henry A. Wallace Building
900 East Grand Avenue
Des Moines, IA 50319
(515) 281-8690

Kansas

Kansas Department of Health and Environment
Bureau of Water Quality
Forbes Field, Building 740
Topeka, KS 66612
(913) 862-9360

Kentucky

Department of Environmental Protection
Division of Water Quality
Fort Boone Plaza
18 Reilly Road
Frankfort, KY 40601
(502) 564-3410

Louisiana
(and EPA Region 6)

Louisiana Department of Environmental Quality
Office of Water Resources
P.O. Box 44091
Baton Rouge, LA 70804
(504) 922-0530

Maine
(and EPA Region 1)

Department of Environmental Protection
Bureau of Water Quality Control
State House Station #17
Augusta, ME 04333
(207) 289-3355

Maryland

Maryland Department of the Environment
201 W. Preston Street
Baltimore, MD 21203
(301) 225-6300

Massachusetts
(and EPA Region 1)

Massachusetts Department of Environmental Protection
Division of Water Pollution Control
One Winter Street
Boston, MA 02108
(617) 292-5673

Michigan

Department of Natural Resources
Water Resources Commission
Water Quality Division
Box 30028
Lansing, MI 48909
(517) 373-1949

Minnesota

Minnesota Pollution Control Agency
Division of Water Pollution Control
520 Lafayette Road, North
St. Paul, MN 55155
(612) 296-7202

Mississippi

Department of Natural Resources
Surface Water Division
P.O. Box 10385
Jackson, MS 39209
(601) 961-5171

Missouri

Department of Natural Resources
Water Quality Program
Jefferson Building - 205 Jefferson St.
Jefferson City, MO 65102
(314) 751-1300

Montana

Department of Health and Environmental Sciences
Water Quality Bureau
Cogswell Building
Helena, MT 59620
(406) 444-2406

Nebraska

Department of Environmental Control
Water Pollution Control Division
State House Station
P.O. Box 94877
Lincoln, NE 68509
(402) 471-2186

Nevada

Department of Conservation and Natural Resources
Water Resources Division
Capitol Complex
201 South Fall Street
Carson City, NV 89710
(702) 885-4380

New Hampshire
(and EPA Region 1)

Department of Health and Welfare
Water Supply and Pollution Control Commission
Health and Welfare Building
6 Hazen Drive
Concord, NH 03301
(603) 271-2458

New Jersey

Department of Environmental Protection
Division of Water Resources
1474 Prospect Street
CN 029
Trenton, NJ 08625
(609) 292-1638

New Mexico
(and EPA Region 6)

New Mexico Health and Environment Department
Surface Water Quality Bureau
1190 St. Francis Drive
Santa Fe, NM 87504-0968
(505) 827-2918

New York

Department of Environmental Conservation
Permit Administrator
50 Wolf Road
Albany, NY 12233

North Carolina

Department of Environment, Health and Natural Resources
Division of Environmental Management
Water Quality Section
P.O. Box 27687
Raleigh, NC 27611
(919) 733-5083

North Dakota

North Dakota State Department of Health
Division of Water Supply and Pollution Control
1200 Missouri Avenue
Bismarck, ND 58501
(701) 224-2345

Northern Mariana Islands

EPA Region 9

Ohio

Ohio Environmental Protection Agency
Waste Water Pollution Control
1800 Watermark Drive
P.O. Box 1049
Columbus, OH 43266-0149
(614) 466-7427

Oklahoma
(and EPA Region 6)

Oklahoma Department of Environmental Quality
Permits and Compliance Division
P.O. Box 53551
1000 Northeast 10th Street
Oklahoma City, OK 73152

Oregon

Department of Environmental Quality
Water Quality Division
522 S.W. Fifth Avenue
P.O. Box 1760
Portland, OR 97207
(503) 229-5324

Pennsylvania

Pennsylvania Department of Environmental Resources
Bureau of Water Quality Management
Fulton Building
200 N. 3rd Street
P.O. Box 2063
Harrisburg, PA 17120
(717) 787-2666

Puerto Rico

Environmental Quality Board
Division of Water Resources
P.O. Box 11488
Santurce, PR 00910-1488
(809) 725-5140

Rhode Island

Department of Environmental Management
Division of Water Resources
209 Cannon Building
75 Davis Street
Providence, RI 02908
(401) 277-2234

South Carolina

Department of Health and Environmental Control
2600 Bull Street
Columbia, SC 29201
(803) 734-4880

South Dakota
(and EPA Region 8)

Department of Water and Natural Resources
Point Source Control Program
Foss Building
120 East Capitol
Pierre, SD 57501
(605) 773-3351

Tennessee

Tennessee Department of Environment and Conservation
Division of Water Quality Control
TERRA Building, 2nd Floor
150 9th Avenue, North
Nashville, TN 37219
(615) 741-3111

Texas
(and EPA Region 6)

Texas Water Commission
P.O. Box 13087, Capitol Station
Austin, TX 78711-3087
(512) 463-8028

Utah

Utah Department of Environmental Quality
Bureau of Water Pollution Control
288 North, 1460 West Street
P.O. Box 16690
Salt Lake City, UT 84116-0690
(801) 538-6146

Vermont

Vermont Department of Environmental Conservation
Water Quality Division
103 South Main Street
Waterbury, VT 05676
(802) 244-5638

Virgin Islands

EPA Region 2

Virginia

State Water Control Board
211 N. Hamilton Street
P.O. Box 11143
Richmond, VA 23230
(804) 257-0056

Washington

Washington Department of Ecology
Office of Water Programs
Mail Stop PV-11
Olympia, WA 98504
(206) 459-6000

West Virginia

West Virgina Department of Natural Resources
Division of Water Resources
1800 washington Street, East
Charleston, WV 25305
(304) 348-2107

Wisconsin

Department of Natural Resources
Bureau of Water Resources and Management
P.O. Box 7921
Madison, WI 53707
(608) 266-2121

Wyoming

Department of Environmental Quality
Water Quality Division
Herschler Building
122 West 25th Street
Cheyenne, WY 82002
(307) 777-7781

Appendix C

SAMPLE POLLUTION PREVENTION PLAN

Double Scoop Ice Cream Company

40 Wonka Drive
Anytown, OK 12345

December 1992

Storm Water Pollution Prevention Plan	
Emergency Contact: **Cheryl Glenn**	Work Phone: **(101) 555-1234**
Title: **Plant Manager**	Emergency Phone: **(101) 555-6929**
Secondary Contact: **Rachel Meyers**	Work Phone: **(101) 555-3923**
Title: **Engineering Supervisor**	Emergency Phone: **(101) 555-6789**
Type of Manufacturer: **Ice Cream Manufacturer**	
Operating Schedule: **8:00 a.m. - 11:30 p.m.**	
Number of Employees: **The plant has 21 employees, including part time staff. Shifts overlap all day.**	
Average Wastewater Discharge: **5,000 gallons per week**	
NPDES Permit Number: **OK1234567**	

POLLUTION PREVENTION TEAM — MEMBER ROSTER	Worksheet #1 Completed by: Cheryl Glenn Title: Plant Manager Date: December 12, 1992

Leader: Cheryl Glenn Title: Plant Manager

Office Phone: (101) 555-1234

Responsibilities: Signatory authority; coordinate all stages of plan development and implementation; coordinate employee training program; Keep all records and ensure reports are submitted.

Members:

(1) Stephen Michaels Title: Production Supervisor

Office Phone: (101) 555-3923

Responsibilities: Note any process changes; help conduct inspections.

(2) Rachel Meyers Title: Engineering Dept. Supervisor

Office Phone: (101) 555-5890

Responsibilities: Responsible for implementing the Preventive Maintenance program; oversee inspections.

(3) Isaac Feldman Title: Maintenance Dept. Supervisor

Office Phone: (101) 555-0482

Responsibilities: Mr. Feldman is the spill response coordinator; Oversees "good housekeeping."

(4) Group Activities Title:

Office Phone:

Responsibilities: Developing the plan elements, chosing storm water management options.

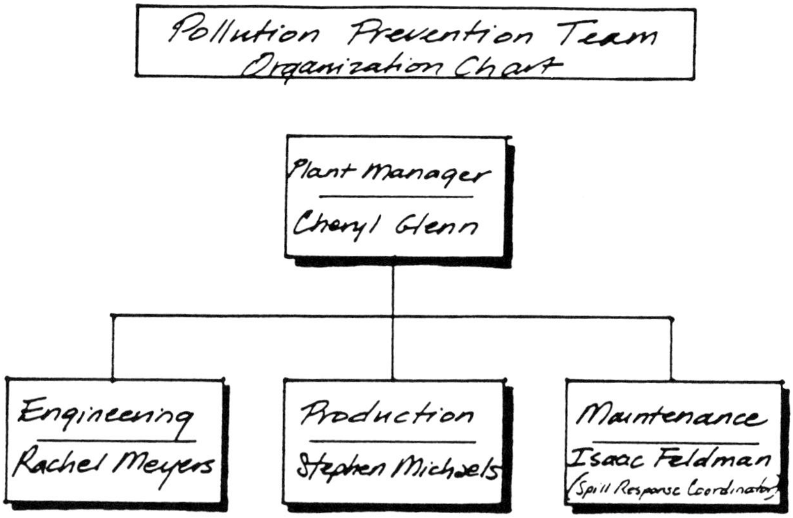
Pollution Prevention Team
Organization Chart
Plant Manager
Cheryl Glenn
Engineering
Rachel Meyers
Production
Stephen Michaels
Maintenance
Isaac Feldman
(Spill Response Coordinator)

Double Scoop Ice Cream Company

Storm Water Pollution Prevention Plan
Comparison with SPCC Plan

Double Scoop Ice Cream Plant has an SPCC plan in operation for its aboveground fuel storage tank. Overlaps are noted below:

- Isaac Feldman is the SPCC Coordinator and reports directly to Cheryl Glenn. He will be the Storm Water Spill Prevention and Response Coordinator.
- A complete description of potential for oil to contaminate storm water discharges including quantity of oil that could be discharged.
- Curbing around aboveground fuel storage tank identified on site map.
- Expanded SPCC schedules and procedures to include Storm Water Pollution Prevention Plan requirements.
- Incorporated SPCC plan training into storm water training programs on spill prevention and response.
- Relevant portions of the SPCC plan will be included in this plan.

DEVELOPING A SITE MAP	Worksheet #2 Completed by: Cheryl Glenn Title: Plant Manager Date: December 12, 1992

Instructions: Draw a map of your site including a footprint of all buildings, structures, paved areas, and parking lots. The information below describes additional elements required by EPA's General Permit (see example maps in Figures 2.3 and 2.4).

EPA's General Permit requires that you indicate the following features on your site map:

- All outfalls and storm water discharges
- Drainage areas of each storm water outfall
- Structural storm water pollution control measures, such as:
  - Flow diversion structures
  - Retention/detention ponds
  - Vegetative swales
  - Sediment traps
- Name of receiving waters (or if through a Municipal Separate Storm Sewer System)
- Locations of exposed significant materials (see Section 2.2.2)
- Locations of past spills and leaks (see Section 2.2.3)
- Locations of high-risk, waste-generating areas and activities common on industrial sites such as:
  - Fueling stations
  - Vehicle/equipment washing and maintenance areas
  - Area for unloading/loading materials
  - Above-ground tanks for liquid storage
  - Industrial waste management areas (landfills, waste piles, treatment plants, disposal areas)
  - Outside storage areas for raw materials, by-products, and finished products
  - Outside manufacturing areas
  - Other areas of concern (specify:__________________)

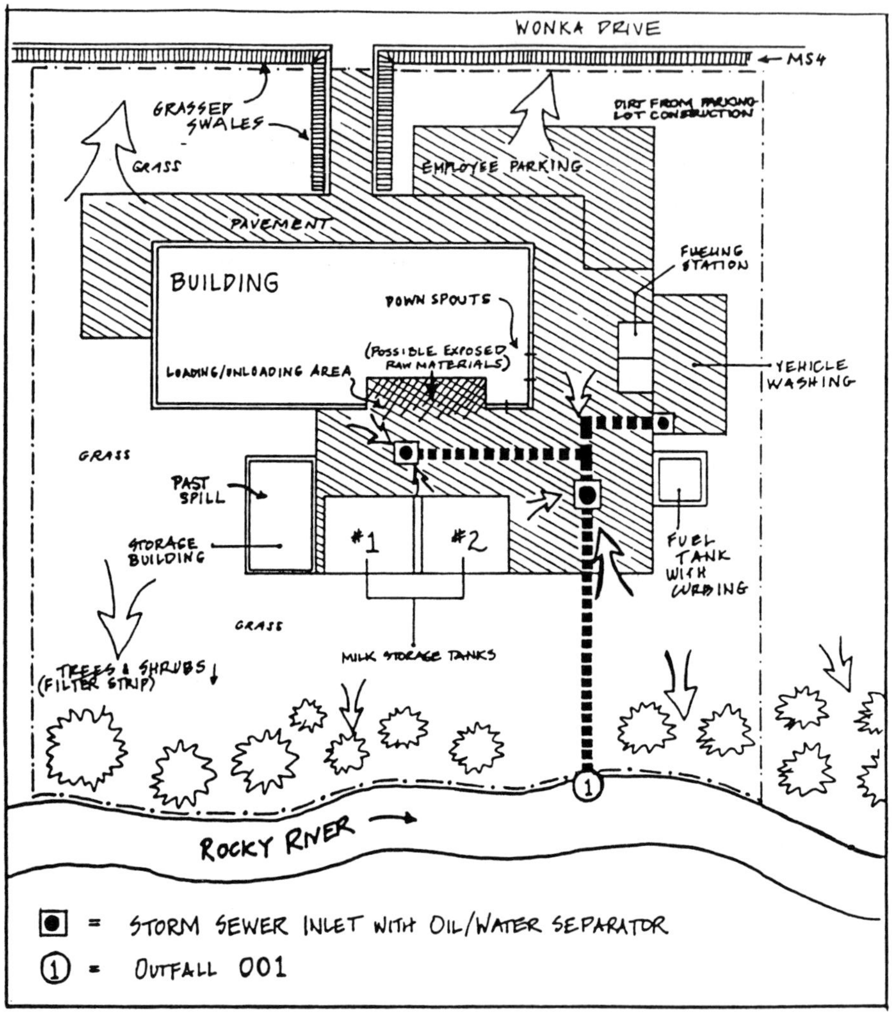

DOUBLE SCOOP ICE CREAM COMPANY

PRE-BMP SITE MAP
MARCH 1, 1993

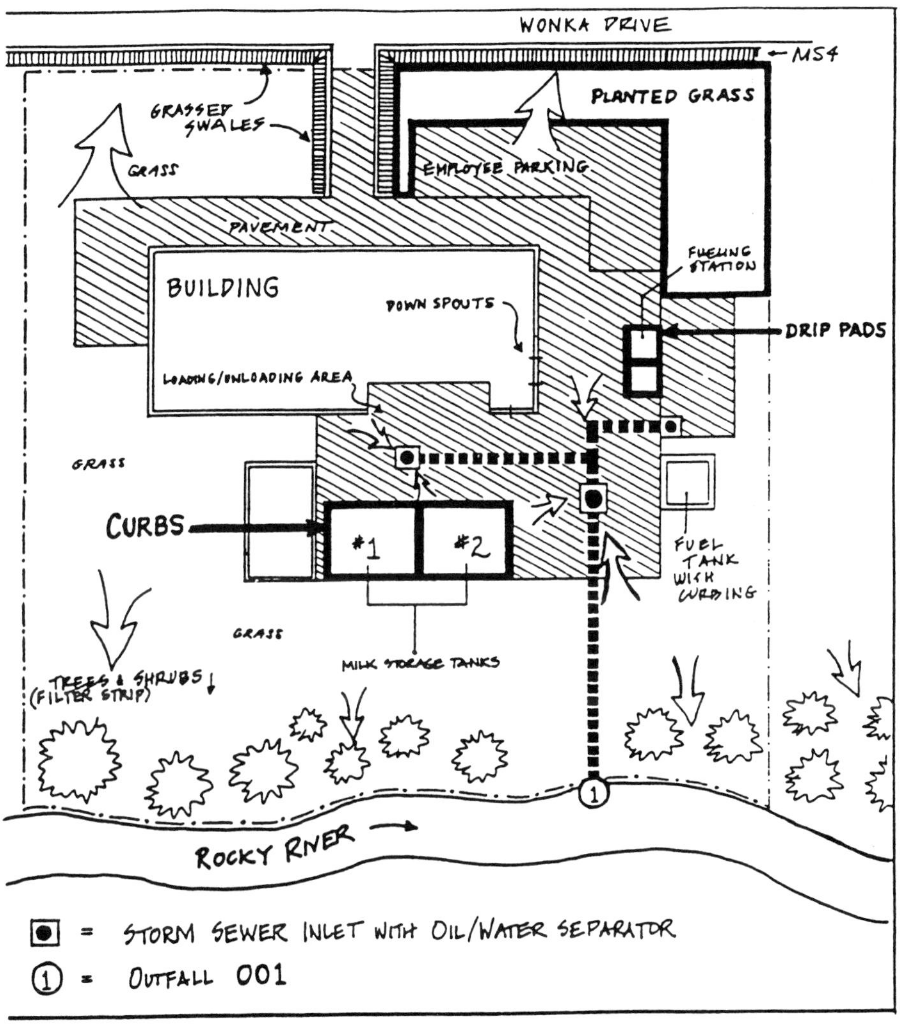

DOUBLE SCOOP ICE CREAM COMPANY

POST-BMP SITE MAP
MARCH 1, 1993

MATERIAL INVENTORY

Worksheet #3
Completed by: Cheryl Glenn
Title: Plant Manager
Date: December 12, 1992

Instructions: List all materials used, stored, or produced onsite. Assess and evaluate these materials for their potential to contribute pollutants to storm water runoff. Also complete Worksheet 3A if the material has been exposed during the last three years.

Material	Purpose/Location	Quantity (units) Used	Produced	Stored	Quantity Exposed in Last 3 Years	Likelihood of contact with storm water. If yes, describe reason.	Past Significant Spill or Leak Yes	No
Butter fat, Milk solids, Whey solids	truck unloading area during transfer to liquid ingredient storage and milk vat storage.	72,600 gal/wk	—	2,000 gal/wk	NO	Truck loading area outside and possible exposure with ruptured tanks.		✓
Corn Syrup, Liquid sugar	Truck unloading area during transfer to sweetener storage.	7,100 gal/wk	—	—	yes	Truck loading area outside with possible exposure as a result of leaking tanks.	✓	
Ice cream	Inside freezers for final product shipping.		35-40,000 Lbs.		NO	NO		✓
Cleansers:								
Granular, Chlorshire-O, H.D.C.-3R, Power Spray-R	Dry cleansers in dry storage area (indoors)	400 lb/wk	—	—	NO	Yes. Possible storage exposure during transfer to dry storage area.		✓

MATERIAL INVENTORY

Page 2

Worksheet #3
Completed by: Cheryl Glenn
Title: Plant Manager
Date: December 12, 1992

Instructions: List all materials used, stored, or produced onsite. Assess and evaluate these materials for their potential to contribute pollutants to storm water runoff. Also complete Worksheet 3A if the material has been exposed during the last three years.

Material	Purpose/Location	Quantity (units)			Quantity Exposed in Last 3 Years	Likelihood of contact with storm water. If yes, describe reason.	Past Significant Spill or Leak	
		Used	Produced	Stored			Yes	No
Cleansers:								
liquid	Cleansers are	100 gal/wk	—	—	NO	Yes-if material tanks		✓
M.R.S.-200-0	stored outside					stored outside.		
Acidize-0	under cover.							
Microsan								
Fuels:								
gasoline	above ground	250 gal/wk	—	—	NO			✓
motor oil	750 gallon storage tank	20 gal/wk	—	—	NO	Yes - possible exposure		✓
						in the event of		
soaps		40 gal/wk	—	—	NO	defective tanks or		✓
detergents						transfer of materials		
						from tanks to containers		

DESCRIPTION OF EXPOSED SIGNIFICANT MATERIAL

Worksheet #3A
Completed by: Cheryl Glenn
Title: Plant Manager
Date: December 12, 1992

Instructions: Based on your material inventory, describe the significant materials that were exposed to storm water during the past three years and/or are currently exposed. For the definition of "significant materials" see Appendix B of the manual.

Description of Exposed Significant Material	Period of Exposure	Quantity Exposed (units)	Location (as indicated on the site map)	Method of Storage or Disposal (e.g., pile, drum, tank)	Description of Material Management Practice (e.g., pile covered, drum sealed)
liquid sugar	1/21/92	10 gal.	Storage building Tank #2	50 gal. tanks (2)	Leak was contained and mopped up. Remainder of liquid sugar transferred to tank that did not have a leaky valve.

LIST OF SIGNIFICANT SPILLS AND LEAKS

Worksheet #4
Completed by: Cheryl Glenn
Title: Plant Manager
Date: December 12, 1992

Directions: Record below all significant spills and significant leaks of toxic or hazardous pollutants that have occurred at the facility in the three years prior to the effective date of the permit.

Definitions: Significant spills include, but are not limited to, releases of oil or hazardous substances in excess of reportable quantities.

1st Year Prior

Date (month/day/year)	Spill	Leak	Location (as indicated on site map)	Description: Type of Material	Description: Quantity	Description: Source, If Known	Description: Reason	Response Procedure: Amount of Material Recovered	Response Procedure: Material No Longer Exposed to Storm Water (True/False)	Preventive Measures Taken
				N	O	N	E			

2nd Year Prior

Date (month/day/year)	Spill	Leak	Location (as indicated on site map)	Description: Type of Material	Description: Quantity	Description: Source, If Known	Description: Reason	Response Procedure: Amount of Material Recovered	Response Procedure: Material No Longer Exposed to Storm Water (True/False)	Preventive Measures Taken
1/21/91		✓	STORAGE BLDG.	LIQUID SUGAR	10 gal.	Tank #2	leaky valve	contained and mopped spill—transferred	true	installing curbing ~~around~~ tank—to be completed → 2/93

3rd Year Prior

Date (month/day/year)	Spill	Leak	Location (as indicated on site map)	Description: Type of Material	Description: Quantity	Description: Source, If Known	Description: Reason	Response Procedure: Amount of Material Recovered	Response Procedure: Material No Longer Exposed to Storm Water (True/False)	Preventive Measures Taken
				N	O	N	E			

NON-STORM WATER DISCHARGE ASSESSMENT AND CERTIFICATION

Worksheet #5
Completed by: Rachel Meyers
Title: Engineering Department Supervisor
Date: 3/1/93

Date of Test or Evaluation	Outfall Directly Observed During the Test (identify as indicated on the site map)	Method Used to Test or Evaluate Discharge	Describe Results from Test for the Presence of Non-Storm Water Discharge	Identify Potential Significant Sources	Name of Person Who Conducted the Test or Evaluation
12/24/92	001	visual inspection	No discharge observed		R. Meyers and S. Goodhope
1/19/93	001	visual inspection	Significant flow; oil	vehicle wash ongoing at time	R. Meyers and S. Goodhope
2/5/93	001	visual inspection	Small amount of discharge observed; clear	suspected to be delayed storm water discharge from storm that occurred 2/1/93	R. Meyers and S. Goodhope
	★See details in attached field notebook.				

CERTIFICATION

I, Cheryl Glenn (responsible corporate official), certify under penalty of law that this document and all attachments were prepared under my direction or supervision in accordance with a system designed to assure that qualified personnel properly gather and evaluate the information submitted. Based on my inquiry of the person or persons who manage the system or those persons directly responsible for gathering the information, the information submitted is, to the best of my knowledge and belief, true, accurate, and complete. I am aware that there are significant penalties for submitting false information, including the possibility of fine and imprisonment for knowing violations.

A. Name & Official Title (type or print): Cheryl Glenn

B. Area Code and Telephone No.: (101) 555-1234

C. Signature: Cheryl Glenn

D. Date Signed: 3/2/93

FIELD NOTEBOOK
For non-storm water discharge inspections

INSPECTION TEAM:
R. Meyers
S. Goodhope

Completed by: Rachel Meyers
Date: 12/24/92
Time: 10:50 am

Time since last rain: 42 hours
Quantity of last rain: 0.12 inches
Flow observed: NO

SIGNATURE: Rachel Meyers

Completed by: Rachel Meyers
Date: 1/19/93
Time: 3:20 pm

Time since last rain: 5 days
Quantity of last rain: 0.5 inches
Flow observed: YES
DESCRIPTION: No odor; clear color (soap suds); oily sheen; some sediment.
Temperature: cold (37.5°F)
Volume: collected ten gallons/minute in buckets
Comments: Vehicle wash ongoing at time of inspection. This was the source of the flow.

SIGNATURE: Rachel Meyers

Completed by: Rachel Meyers
Date: 2/5/93
Time: 12:15 pm.

Time since last rain: 96 hours
Quantity of last rain: 2.5 inches
Flow observed: YES
DESCRIPTION: No odor; clear; some sediments; few small pieces of paper (trash)
Temperature: cold (42.3°F)
Volume: Collected one gallon in 5 minutes.
Comments: We suspect that the flow was left over from storm that occurred on 2/1/93 (4 days ago)

SIGNATURE: Rachel Meyers

Double Scoop Ice Cream Company

Site Assessment Inspection

February 10, 1993

Evaluate the site for pollutants.

There are five areas where material handling and storage activities take place.

- The storage building contains tanks of corn syrup, liquid sugar, and the granular cleansers. The tanks were examined for possible leaks. We found that the valve on the liquid sugar tank #2 was faulty and had leaked approximately 10 gallons of liquid sugar. Although this leak occurred on 1/21/92, the faulty valve was not discovered until now. All other tanks are secure. Areas around the tanks were swept clean to determine if leaks or spills were prevalent.

- The milk storage tanks were then examined for leaks or exposure. Upon closer examination, it was found that the number 1 tank was leaking a small amount of milk to the drainage system. This leak may be the reason for the high concentration of biochemical oxygen demand found in the sample taken from the storm water discharge. The tank was temporarily fixed to ensure that no further contamination would result. A replacement tank was ordered on February 6, 1993, and was expected to arrive within 5 business days. The milk storage tanks shall be examined on a daily basis to further prevent possible exposure to the storm water collection system and receiving stream.

- We inspected the fueling station to see if there were any leaks. The general area surrounding the fueling station was clean but we observed that gasoline and motor oil falls during fueling. In accordance with standard operating conditions, facility personnel hose down the area during vehicle washing and the drain is connected to the storm sewer. We detected this connection on 1/19/93 during one of the non-storm water discharge assessment visual inspections. Since this discharge is not allowed under our general permit, we are in the process of submitting a separate permit application specifically for the discharge of vehicle wash water.

- We examined the fueling station which is adjacent to the vehicle washing area. Vehicle washing cleaners are used here and any empty or open containers were removed from the area.

- We next looked at the loading and unloading docks where raw materials and various cleansers are delivered. The transfer of goods from incoming trucks to storage areas is a source of pollution. Although no problems were noticed, the pollution prevention team has developed a spill prevention and response plan to clean up spills quickly and report them if necessary.

- The last area we inspected was the runoff field below the employee parking lot. Here we noticed a significant amount of erosion resulting from recent construction to expand the parking lot.

Describe existing management practices.

Grass was lightly planted around the parking lot after recent construction. The fuel storage tank has curbing around it in accordance with our SPCC plan. Also, the maintenance crew regularly picks up trash and empty containers from around the storage tanks, loading and unloading areas, and the vehicle washing areas. Used oils are collected in containers and taken to a recycling facility. In addition, we installed two oil/water separators at the drains into our underground storm sewer leading to the Rocky River. These separators are indicated on the site map.

Double Scoop Ice Cream Company

Existing Monitoring Data

Although our NPDES permit for process wastewater does not require storm water sampling, we sampled our storm water on one occasion in response to a questionnaire we received from the National Association of Ice Cream Makers. They were collecting information to submit as part of their comments on EPA's proposed general permit.

Date of Sampling	8/30/91
Outfall Sampled	001
Type of Storm	1 inch light rainfall (lasted 2 days)
Type of Samples	Grab samples taken during first hour of flow

Data		
Parameter	**Quantity**	**Sample Type**
BOD	250 mg/l	Grab
TSS	100 mg/l	Grab
pH	7.2 s.u.	Grab
Oil and grease	5.0 mg/l	Grab

Based upon the high concentration of BOD in the storm water samples collected, pollution prevention team is considering possible potential sources of BOD. We will look at storage areas housing butter fat, milk, and whey solids tanks.

Double Scoop Ice Cream Company

Summary of Pollutant Sources

March 5, 1993

Based on the site assessment inspection conducted on 12/1/92, the pollution prevention team identified four potential sources of pollutants:

- Oil and grease stains on the pavement in the fueling area indicate oil and grease may be picked up by storm water draining to the storm sewer. This area drains into the storm sewer leading to the Rocky River.
- Sediment and erosion potential in the field below the employee parking lot because of thinly planted grass.
- Potential for spills or leaks from liquid storage tanks, including the fuel storage tank, based on a spill that occurred on 1/21/92 and the leak that was detected in the milk storage tank. These pollutants would drain into the piped outfall into the Rocky River.
- Use of a toxic cleaning agent may result in a pollution problem if handled improperly.

Double Scoop Ice Cream Company

Description of Storm Water Management Measures Taken
Based on Site Assessment Phase

March 5, 1993

These measures correspond to the pollutant sources identified on the preceding page.

Oil and grease from fueling area.

We installed drip pads around the fuel pumps to pick up spilled gas and oil during truck refueling. These will be inspected regularly to make sure they are working well.

Sediment and erosion in the field below the employee parking lot.

We planted grass in this area to reduce potential for erosion.

Leaks/spills from liquid storage tanks.

We are in the process of installing curbing around the outdoor liquid storage tanks that will contain the volume of he largest tank in case a spill should occur. The spill response team has developed procedures to clean up this area should a spill occur. We are incorporating spill response procedures from our SPCC plan.

Toxic cleaning agent.

We have discontinued the use of this agent and are replacing it with a non-toxic cleaning agent.

POLLUTANT SOURCE IDENTIFICATION
(Section 2.2.6)

Worksheet #7
Completed by: Cheryl Glenn
Title: Plant Manager
Date: 3/5/93

Instructions: List all identified storm water pollutant sources and describe existing management practices that address those sources. In the third column, list BMP options that can be incorporated into the plan to address remaining sources of pollutants.

Storm Water Pollutant Sources	Existing Management Practices	Description of New BMP Options
1. Oil and grease on pavement in fueling area	Oil and water separators installed in storm water drain	Install drip pads
2. Erosion in field below employee parking lot	Planted some grass after construction; grassed swales along Wonka Drive	Plant more grass
3. Potential for spills from liquid storage tanks (leak detected in milk tank #1; past spill on 1/21/92)	Curbing around fuel storage tank (see SPCC plan)	Replace milk tank #1; replace valve on liquid sugar tank #2; install curbing around other outside tanks; spill prevention response plans; inspection
4. Use of toxic cleaning agent		Use non-toxic cleaning agent
5. Trash in loading/unloading fueling areas	Regular trash pickup (daily) by maintenance crew; collect and recycle used oil.	Train staff in good housekeeping practices
6.		
7.		
8.		
9.		
10.		

BMP IDENTIFICATION
(Section 2.3.1)

Worksheet #7a
Completed by: Cheryl Glenn
Title: Plant Manager
Date: 3/5/93

Instructions: Describe the Best Management Practices that you have selected to include in your plan. For each of the baseline BMPs, describe actions that will be incorporated into facility operations. Also describe any additional BMPs [activity-specific (Chapter 3) and site-specific BMPs (Chapter 4)] that you have selected. Attach additional sheets if necessary.

BMPs	Brief Description of Activities
Good Housekeeping	Collect and recycle used oil; regular trash pick up; train staff in basic clean up procedures (sweeping loading & unloading areas, etc.)
Preventive Maintenance	Daily inspection of outside milk tanks; replace faulty valve on sugar tank #2; replace leaking milk tank #1
Inspections	Daily inspection of outside milk tanks; bi-monthly inspections of drip pads, curbing, loading/unloading areas, grassed areas, drainage system.
Spill Prevention Response	Install curbing around outside liquid storage tanks; fuel tank has curbing; install drip pads at fueling station.
Sediment and Erosion Control	Plant grass around new parking area.
Management of Runoff	Grassed swales along Wonka Drive, (2) oil/water separators in storm drain system
Additional BMPs (Activity-specific and Site-specific)	Order non-toxic cleaning agent.

Double Scoop Ice Cream Company

Employee Training Program

Who:

Line Workers
Maintenance Crew
Shipping and Receiving Crew

When:

Employee meetings held the first Monday of each month to discuss:

- Any environmental/health and safety incidents
- Upcoming training sessions
- Brief reminders on good housekeeping, spill prevention and response procedures, and material handling practices
- Announce any changes to the plan
- Announce any new management practices

In-depth pollution prevention training for new employees

Refresher courses held every 6 months (October and March) addressing:

- Good housekeeping
- Spill prevention and response procedures
- Materials handling and storage

Employee Training Program Topics:

Good Housekeeping

- Review and demonstrate basic cleanup (sweeping and vacuuming) procedures.
- Clearly indicate proper disposal locations.
- Post signs in materials handling areas reminding staff of good housekeeping procedures.
- Be sure employees know where routine clean-up equipment is located.

Spill Prevention and Response

- Clearly identify potential spill areas and drainage routes
- Familiarize employees with past spill events -- why they happened and the environmental impact (use slides)
- Post warning signs in spill areas with emergency contacts and telephone numbers
- Introduce Isaac Feldman as the Spill Response Coordinator and introduce his "team"
- Drill on spill clean-up procedures
- Post the locations of spill clean-up equipment and the persons responsible for operating the equipment

Materials Handling and Storage

- Be sure employees are aware which materials are hazardous and where those materials are stored
- Point out container labels
- Tell employees to use the oldest materials first
- Explain recycling practices
- Demonstrate how valves are tightly closed and how drums should be sealed
- Show how to fuel vehicles and avoid "topping off"

IMPLEMENTATION
(Section 2.4.1)

Worksheet #8
Completed by: Cheryl Glenn
Title: Plant Manager
Date: 3/30/93

Instructions: Develop a schedule for implementing each BMP. Provide a brief description of each BMP, the steps necessary to implement the BMP (i.e., any construction or design), the schedule for completing those steps (list dates) and the person(s) responsible for implementation.

BMPs	Description of Action(s) Required for Implementation	Scheduled Completion Date(s) for Req'd. Action	Person Responsible for Action	Notes
Good Housekeeping	1. Develop training program	3/10/93	Glenn	
	2. Conduct training	6/1/93	Glenn	
	3.			
Preventive Maintenance	1. Replace valve on sugar tank #2	3/1/93	Feldman	
	2. Install new milk tank #2	2/15/93	Feldman	
	3.			
Inspections	1. Develop inspections schedule	4/1/93	Glenn	
	2.			
	3.			
Spill Prevention and Response	1. Install curbing around milk storage tanks	4/30/93	Meyers	
	2. Install drip pads	4/1/93	Feldman	
	3. Develop/Implement Spill Prevention/Response training	4/1/93 - DEVELOP 6/1/93 - TRAIN	Feldman	
Sediment and Erosion Control	1. Plant grass around parking area	4/15/93	Feldman	
	2.			
	3.			
Management of Runoff	1. BMPs already in place			
	2.			
	3.			
Additional BMPs (Actively-specific and site-specific)	1. Substitute non-toxic cleansing agent	2/28/93	Michaels	
	2.			
	3.			

EMPLOYEE TRAINING
(Section 2.4.2)

Worksheet #9
Completed by: Cheryl Glenn
Title: Plant Manager
Date: 3/2/93

Instructions: Describe the employee training program for your facility below. The program should, at a minimum, address spill prevention and response, good housekeeping, and material management practices. Provide a schedule for the training program and list the employees who attend training sessions.

Training Topics	Brief Description of Training Program/Materials (e.g., film, newsletter course)	Schedule for Training (list dates)	Attendees
Spill Prevention and Response	locate spill areas by signs; drill spill response procedures; show slides of past spills.	October/March	Maintenance/ shipping & receiving
Good Housekeeping	Demonstration; post signs at disposal sites.	October/March	Maintenance/ shipping & receiving
Material Management Practices	Introduce hazardous materials labels; discuss recycling.	October/March	Lineworkers/shipping and receiving/maintenance
Other Topics	Environmental/health incidents; Reminders of pollution prevention plan issues.	1st Monday of each month	All employees.

Index

G

N

O

P

S